MINDLESS

ALSO BY SIMON HEAD

The New Ruthless Economy

MINDLESS

Why Smarter Machines
Are Making Dumber Humans

SIMON HEAD

BASIC BOOKS
A Member of the Perseus Books Group
New York

Published by Basic Books,

A Member of the Perseus Books Group

Books published by Basic Books are available at special discounts for bulk purchases in the United States by corporations, institutions, and other organizations. For more information, please contact the Special Markets Department at the Perseus Books Group, 2300 Chestnut Street, Suite 200, Philadelphia, PA 19103, or call (800) 810-4145, ext. 5000, or e-mail special.markets@perseusbooks.com.

Designed by Jeff Williams

Library of Congress Cataloging-in-Publication Data
Head, Simon.
  Mindless : why smarter machines are making dumber humans / Simon Head.
  pages cm
Includes bibliographical references and index.
  ISBN 978-0-465-01844-4 (hardback) — ISBN 978-0-465-06974-3 (e-book) 1. Technology—Social aspects. 2. Business—Data processing—Psychological aspects. 3. Industries—Techno-logical innovations—Psychological aspects. 4.—Mental efficiency. 5. Knowledge management. I. Title.

T14.5.H445 2013
303.48'3—dc23
                                                                            2013041878

10 9 8 7 6 5 4 3 2 1

For Dee and Nelson Aldrich

CONTENTS

INTRODUCTION

Toward a New Industrial State

ALTHOUGH INEQUALITY OF INCOME AND WEALTH IN AMERICA HAS been growing steadily for the past forty years, it was with the Wall Street crash of 2007–2008 that this disparity took on lurid, visible form with the contrasting fortunes of the winners and losers. On the winning side, with their big bonuses, were many Wall Streeters who themselves bore responsibility for the crash. On the losing side were victims of the crash on Main Street, burdened with high unemployment, crushing personal debt, falling real wages, and shrinking personal wealth propelled by housing foreclosures.

By grim coincidence, detailed statistical evidence of how extreme American inequality had become also appeared during the crisis year of 2007. The data revealed the great good fortune of the super rich—of the richest 1 percent, 0.1 percent, and even the richest 0.01 percent of Americans. The share of total income of the top 1 percent rose from 8 percent in 1974 to 18 percent in 2007 and from

9 percent to 23.5 percent if capital gains and income from investments were included. The equivalent share of the richest 0.1 percent of Americans rose from 2.7 to 12.3 percent and the share of the very richest, the top 0.01 percent, from less than 1 percent to 6 percent during the same period.[1]

The reverse side of this massive concentration of income and wealth at the earnings pinnacle—unprecedented since the pre-1914 Gilded Age—is the stagnation or fall in the real incomes of virtually everybody else. The growth of median annual earnings of most Americans has been spectacularly weak, irrespective of educational attainment. Between 1980 and 2006 the median annual earnings of fully employed entry-level workers between the ages of twenty-five and thirty-four with a bachelor's degree or higher increased by just $1,000 in constant 2006 dollars, from $44,000 to $45,000, for a total percentage increase of just 2.27 percent over a twenty-six-year period, or an increase of less than 0.1 percent a year. The real earnings of those with some college education but with less than a four-year bachelor's degree fell by $5,300 over the same period, or a percentage fall of 14.5 percent. For those with a high school diploma or equivalent, the comparable figure was a fall of $5,200 in constant dollars, for a total percentage fall of 15.3 percent.[2]

Unless soon reversed, these decades of income stagnation or decline for the majority threaten something fundamental to American identity that for more than two centuries has set the United States apart from its old European mother countries: the confidence of most Americans that through education and hard work, they can overcome the barriers of birth and inheritance and rise as far as their talents will take them. This confidence is draining away as the barriers of American class strengthen, shrinking the life prospects of what may now be a majority of Americans and including much of the middle class among the newly disadvantaged.

As its title suggests, this book will look at the role of information technology (IT) as a driver of this inequality. By making us dumber, smart machines also diminish our earning power. But the machines that do this are not the automating, stand-alone machine tools of the 1950s, or even the stand-alone mainframes of the 1960s and 1970s, but vast networks of computers joined by software systems and the Internet, with the power to manage the affairs of giant global corporations *and* to drill down and micromanage the work of their single employees or teams of employees. There now exist in the US economy of the new century these very powerful agents of industrialization, known as *Computer Business Systems (CBSs)*, that bring the disciplines of industrialism to an economic space that extends far beyond the factories and construction sites of the industrial economy of the machine age: to wholesale and retail, financial services, secondary and higher education, health care, *"customer relations management"* and *"human resource management (HRM),"* public administration, corporate management at all levels save the highest, and even the fighting of America's wars.

CBSs are being pushed by business academics, especially at the Massachusetts Institute of Technology (MIT), management consultants such as Accenture and Gartner, and IT companies such as SAP, IBM, and Oracle, and embraced by corporations for their efficiency. But they are not well understood beyond these specialist communities engaged in their creation, marketing, and servicing. These systems are today rather as black holes once were before black holes were fully discovered. Astrophysicists knew that there were things out there in the cosmos exerting a gigantic gravitational pull over everything that came into contact with them, but they did not yet know exactly what these things were. CBSs are the semidiscovered black holes of the contemporary economy.

One measure of their obscurity is that there is no generally accepted name for them. Some of the most influential economists doing work in the field call them *Computer Business Systems,* and I am following their example here. But they have also been known as *Enterprise Systems* and by several other names and activities closely associated with them at various stages of their history: *Business Process Reengineering (BPR)* in the early and midnineties, *Enterprise Resource Planning (ERP)* in the mid- and late nineties, and the *Balanced Scorecard (BSC)* throughout the 1990s and into the new century. Yet despite this obscurity and lack of a fixed identity, evidence occasionally surfaces showing how much the corporate sector relies on these systems and how heavily it has invested in them.

In 1995 a report commissioned by the Big Three accounting companies reported that 75 to 80 percent of America's largest companies were engaged in Business Process Reengineering and "would be increasing their commitment to it over the next few years."[3] A 2001 report cited by economists Eric Brynjolffson of MIT and Andrew McAfee of the Harvard Business School estimated that in 2001, investment in ERP systems accounted for 75 percent of all US corporate IT investment. Typically, the introduction of CBSs costs large corporations hundreds of millions of dollars, and their full implementation can take years to achieve. In the early 2000s, Cisco Systems budgeted $200 million to be spent over three years for its CBS upgrade.[4] This management "giantism" is also a global phenomenon. In China leading American management consultants are devoting much of their effort to the introduction of SAP systems to Chinese state enterprises undergoing privatization.[5]

The human side of this new industrialism can easily get lost in the abstract, theoretical world of macroeconomics and management science. In the first machine age the working class occupied a world apart, tethered to factories and assembly lines and bearing the full rigors of industrialism. In the new machine age, the working class can be all of us. The new industrialism has pushed out from its old heartland in manufacturing to encompass much of the service economy, and it has also pushed upward in the occupational hierarchy to include much of the professional and administrative middle class: physicians as well as call-center agents; teachers, academics, and publishers as well as "associates" at Walmart and Amazon; bank loan officers and middle managers as well fast food workers.

In the first machine age, the primordial conflict was not only about wages and benefits but also about the pace of work, the speed at which the automatic machine and the assembly line would run, and so the rate at which human as well as physical capital would be depleted. With the coming of the networked computer with monitoring software attached, industrial regimes of quantification, targeting, and control now pervade the white-collar world: how many patients, litigants, customers with complaints, students with theses, and future home owners with mortgage applications have been processed or billed per day or week, and how many *should* be processed or billed, because the digital white-collar line is subject to speedup no less than its factory counterpart?

White-collar professionals subject to relentless targeting and speedup have to wonder whether they, like shop-floor employees at Walmart and Amazon, are being worked and worked until they too become depleted as *human resources (HR),* victims of burnout, then "let go," to be thrown onto the human slag heap just like the nineteenth-century proletarians of Émile Zola's great novel about the coal miners of northern France, *Germinal.* In the first machine age, the relations between men and machines were on display in the operations of the factory floor. The abuses that took place were visible to the outside world, the raw material of radicalism and reform.

In the new machine age, the workings of the white-collar line are hidden in the innards of servers and software systems. They are also cloaked in the mystique and prestige of *science* and *high technology*. They now need to be brought into the open.

This is the production world of IT, which leaves behind Steve Job's lustrous and indulgent kingdom of iPods and iPads and opens up an austere, puritan republic in which the relationship between IT and its users is turned on its head. In the Steve Jobs world, the products of IT are our servants and we have the freedom to do what we want with them (though businesses, for their own purposes and profit, closely watch how we exercise this freedom). On the production side of IT, the relationship is transformed and the systems dominate. They enforce the rules that determine how work should be done and with a power and speed unthinkable in the predigital age. But although the systems enforce the rules, they do not make them; they have no will of their own. The rules are the work of a number of interested parties: the senior executives who know broadly what they want the rules to look like, the system providers such as IBM and SAP who supply products whose designs are close to what the executives want, and the corporations' own in-house designers who can tweak the purchased products to account for local needs.

CBSs are amalgams of different technologies that are pulled together to perform highly complex tasks in the control and monitoring of businesses, including their employees. The technologies of the Internet are critical to CBSs because they provide the foundation for computer networks that can link the workstation of every employee or group of employees within an organization to that of every other, irrespective of location and status—from a chief executive officer (CEO) in New York to a group of claims processors in Bangalore, India.

Products known as "data warehouses" and "data marts" are also critical to the CBS control regime. Data warehouses contain the gigantic quantities of information needed to store data on millions of transactions performed daily by tens of thousands of employees—the raw material of the system. Data marts "cleanse" and order this data so that it can be used to evaluate performance in real time and in line with matrices established by management. Once data warehouses and data marts are fused with the monitoring capabilities of CBSs, then the building blocks of a very powerful system of workplace control are in place.

Most CBSs also contain a third critical element: expert systems that mimic human intelligence in performing the cognitive tasks that are integral to the business processes to be managed by the system. Their presence within the system is essential if complex interactions between humans, as in health care, higher education, customer service, and human resource management, are to be fully subject to the industrial disciplines of measurement, standardization, and speed. The most notorious example of such industrialization via expert systems is their use by health maintenance organization case managers to rule on the treatments that patients should or should not receive from their physicians. A doctor may send in a bill for treatment, but the HMO may refuse to pay it because the treatment did not conform to the HMO's "best practice" as defined by the HMO's own medical experts and as embedded in the rules of the system.

THERE ARE PRECEDENTS in American business history for this pulling together of technologies to form a single technology, performing highly complex tasks. It is what Henry Ford achieved with the technologies of mass production at his Highland Park and River Rouge plants during the second and third decades of the twentieth century. The Rouge plant in particular was for its time a miracle of technology integration, fusing the activities of steel mill, stamping plant, machine shop, and assembly line, transforming the raw materials of iron and steel entering the Rouge at one end into the finished and tested Model T coming out at the other.

The Ford regime is illuminating in another way, because it provides a conceptual framework that makes sense of today's CBSs. This conceptual framework pivots upon a single and modest word, *process,* a word that is nonetheless omnipresent and dominant in the contemporary literature of American business schools, management consultants, corporate mission statements, and "system providers" such as IBM, SAP, and Oracle. Modest it may be, but *process* probably carries more historical baggage than any other single word in the entire corporate vocabulary.

Much of this baggage dates from the mass-production regimes of Ford's own time. Ford defined mass production as "the focusing upon a manufacturing project of power, accuracy, economy and speed,"[6] and these were the paramount characteristics of the processes of automobile production in the Ford plants: the progress of the embryonic Model T along the way stations of production from steel mill to testing station, always following a rigorously timed and standardized sequence of operations.

One of the central distinctions in the sociology of work is between "process" and "practice." *Process* we are already familiar with; it refers to a series of operations and how they relate to one another. *Practice,* on the other hand, refers to the activities that can inhabit each operation in the process and especially to the accumulation of tacit knowledge and skill that employees bring to bear in order to perform well such embedded tasks. In the mass-production regime perfected by Ford, the distinction between "practice" and "process" withers away. "Process" reaches down from the commanding heights, pushes "practice" aside, and extends its domain to the most humble activities in the plant. Thus, in the Ford plants, there was not only the process of making the Model T from steel mill to testing station, but also the process of assembling engine pistons and rods where time and motion studies were applied to eliminate four hours' worth of walking time in the assembler's daily routine.[7]

There was also a second axis of process on display at the Ford plants that, although less resonant in the business history of the twentieth century than the assembly line, has been no less central to the working of the mass-production model. Managers were needed to ensure that the huge, sprawling mechanism of the plant, with its myriad processes both macro and micro, was kept running on an even keel and did not dissolve into chaos. There had to be a continuous flow of information arising from the shop-floor processes, traveling upward through layers of management, conveying to senior managers that processes were or were not running as they should and with production targets being met or not met.

The best account we have of these turn-of-the-century management processes is found in Alfred Dupont Chandler's *Visible Hand: The Managerial Revolution in American Business*,[8] one of the very few great books yet written about management. One might say that whereas both the macro and the micro processes of production were *horizontal* in the sense that their constituent operations followed one another in a precisely calibrated sequence, the processes of management were *vertical* because they consisted of an upward flow of information that rose from the shop floor through layers of management, eventually reaching the corporate pinnacle.

The ubiquity of the word *process* in the contemporary American management literature points both to the descent of today's processes from those of the Ford era and to the differences between the two generations of process. Yet these differences, overwhelmingly bound up with the role of IT in modern-day process, accentuate aspects of process that are usually thought of as belonging to the industrial rather than the postindustrial era: the speed of processes, their standardization, and their susceptibility to timing and control from above. Such tightening and acceleration of contemporary process through IT are evident both in the case of horizontal assembly-line processes, especially with their transfer from the blue-collar to the white-collar world, and in the vertical management processes that in their contemporary incarnation I will call *Corporate Panoptics (CP)*. In the early twenty-first century, the chief redoubt of processes both horizontal and vertical has been the Computer Business System.

There is also a critical difference between CBSs and all the other production systems that have featured prominently in the history of capitalism during the past 250 years—beginning with Adam Smith's description in *The Wealth of Nations* of an eighteenth-century pin factory and continuing with Marx's account of a mid-nineteenth-century English textile mill in volume 1 of *Das Kapital*, then the early description of the Ford system by Horace Arnold, influential in its time, and then most recently the account of Japanese lean production in the auto industry by Womack, Jones, and Roos in *The Machine That Changed the World*.[9]

With these production regimes of the machine age, the systems took on visible forms in ways that could illuminate, often dramatically, the interaction between men and machines. In *The Wealth of Nations*, Adam Smith gives a vivid sense of the pin makers as proto assembly-line workers, each performing a micro task of pin manufacture. Marx notoriously never entered a factory, but his harrowing account of the exploitation of child labor in mid-nineteenth-century English textile mills drew on the evidence of the official factory inspectors who did make visits and whose reports eventually led to the outlawing of child labor in English factories.

The assembly line has been a dominant image of the machine age because the line and its workforce could be visited, watched, photographed, and even dramatized in the movies—notably by Chaplin in *Modern Times* (1936). But what are the visual manifestations of CBSs—a concrete blockhouse somewhere in New Jersey housing the huge servers needed to handle the gigantic quantities of information yielded by the systems, or employees staring at rows of flickering computer screens, receiving their instructions online and then keywording in their responses or, if working in call centers, speaking to customers on the telephone? This visual poverty elevates the importance of the trade literature on CBSs put out by their leading creators—SAP, IBM, and Oracle—as primary sources about what the systems are and how they work.

THIS BOOK OPENS up the largely hidden world of CBSs and explores the ideas and practices of the corporations, consultants, and management theorists who sustain them. This is a missing piece of the economic jigsaw whose absence detracts significantly from our understanding of the US economy at a time when its growing inequalities of income, wealth, and power threaten its social and political well-being as nothing has since the Great Depression. There are explanations for this malaise that, on the face of it, have little or nothing to do with CBSs and the production side of IT. Among them are the displacement of much US manufacturing to the developing world; the shift of political power in favor of business, leveraged by business to skew the distribution of income and wealth in its interest; and the deterioration in US education at all levels that leaves a growing percentage of the labor force without the skills to hold down well-paying jobs in the "knowledge economy" or to compete with the tens of millions entering the global labor force, especially in East Asia.

But the "IT question" as defined here can both challenge and amplify these explanations. Can, for example, the overseas sourcing of manufacturing really be an adequate explanation for the US economic malaise when more than 80 percent of the US labor force is now employed in service industries, which for the most part are not in direct competition with the developing world and where the impact of white-collar industrialization has been especially severe? Then, turning to the US workplace itself, would the top management of US corporations have been so successful in skewing the distribution of corporate profits in their own favor if the workforce really had been empowered by information technology as "knowledge workers" in a "knowledge economy," as management gurus such as Peter Drucker confidently predicted twenty years ago?[10] And is improved education at the high school, vocational, or even college level really the golden key to a world of high-paying, secure employment if in fact Computer Business Systems are being used to marginalize employee knowledge and experience and where employee autonomy is under siege from ever more intrusive forms of monitoring and control?

The emerging relationship between technology and work in the US economy of the late twentieth and early twenty-first centuries suggests that the corporate sector is relying on information technology *both* to simplify and accelerate the processes of business output, and so increase the output of labor, *and* to deskill labor, diminish its role, and so weaken its earning power. The widening gap between the growth of labor's output and its real earnings is the desired outcome of this regime. When the output of labor rises and its real earnings stagnate or decline, then, other things being equal, the cost of labor per unit of output will fall and profits will rise.

From a corporate perspective, this is a good outcome, and especially with the compensation of top management so frequently linked to the corporate stock price, which will tend to rise with profits. But there is an identity and equivalence of basic economics that this project overlooks. Producers are also consumers, and by denying employee-producers the rewards of their increased productivity, the architects of the wages-productivity gap have also laid siege to the consumers' republic and so undermined the US economy's single most powerful engine of demand and growth. Consumers had been relying on debt to keep their consumption afloat in the face of stagnant real earnings, but this remedy, like the housing bubble itself, could not endure and indeed ended with the financial crisis of 2007–2008.

In explaining why the recovery has been so weak and why it is having to keep interest rates so low and for so long, the Federal Reserve has placed a heavy emphasis on the poor financial condition of consumers and their inability to relaunch the economy with their spending, constrained by high unemployment, zero income growth, lower housing wealth, and tight credit.[11] What the Fed does not acknowledge is that the eclipse of consumers is simply the reverse side of their eclipse as producers and that this has taken place as part of an economy-wide business plan.

1

INSIDE THE BELLY OF THE BEAST

IN THIS CHAPTER WE WILL RELY ON THE CBS PRODUCT MANUALS to travel as far as we can into their esoteric world. The obscurity of CBSs, their complexity, and their visual poverty elevate the importance of these manuals as sources about what they are and how they work. Trying to understand the systems without these texts is like trying to climb a Himalayan peak without a guide. In an age of managerial hegemony, it might be thought difficult to find a substantial bibliography of such primary sources concerned with one very significant aspect of CBSs—what its like to be at their receiving end as employees. But a copious bibliography of such sources does exist, and it comes not from labor unions, progressive think tanks, and least of all from the bowdlerized texts of management gurus such as Michael Hammer and James Champey of reengineering fame. It comes from the texts of the IT corporations that make and market the systems themselves.

Foremost among these are IBM, Oracle, and the German corporation SAP, as well as Scheer AG of Saarbrücken, Germany, a *mittelstand* software company that has had a strong and enduring influence on SAP, the world leader by market share for CBSs. Their product manuals between them illuminate with engineering thoroughness whole continents of the CBS world uncharted by the management gurus.[1] Running to five-hundred-plus pages in the case of IBM's Red Books, they are texts written by engineers for engineers, and, as so often happens when technicians turn inward and address one another in their trade literature, they say things about their products that they would not say when facing outward and addressing a wider audience.

The texts rely heavily on an abstract, quasi-scientific language that is a strong deterrent to anyone from beyond the specialist CBS communities wanting to read them. The documents speak of business events and occurrences, critical business situations, process instances and flows, process improvement metrics, and event-driven process chains (EPCs). The CBS engineers use this disembodied language in part because their products are designed for use throughout the economy, and so the language of explanation must be abstract and general. To use a language identified with any one particular segment of the economy, such as manufacturing, would be to imply that there were other segments such as financial services or health care where the systems could not be used.

Behind this *langue de bois* of digital managerialism lurks something truly transformative. The objects of management are no longer flesh-and-blood humans but their electronic representations. We have become the numbers, coded words, cones, squares, and triangles that represent us on digital screens. The human-contact side of management—the tasks of explanation, persuasion, and justification—fades away as workplace rules and procedures become texts showing up on employees' computer screens, with the whole apparatus of monitoring and control instantly recalibrated to accommodate the new metrics.

With the latest generation of CBSs, this control regime has reached far beyond the systems' original base in manufacturing to include virtually the entire service economy, so not only service sectors that are low skilled and labor intensive such as the retail economy with Amazon and Walmart to the fore but also sectors that are skill intensive and the preserve of professionals, such as hospitals and clinics, university lecture and seminar rooms, the offices of banks, insurance companies, government departments, and the laboratories of human resource management.

Early in its main product manual for its Websphere Business Monitor V6.1, IBM stakes out its claim to these professional, white-collar workplaces. The manual describes how its control technologies empower "financial institutions to track and manage loans processes in real time," enable a "government agency to gain visibility into the operations of a social service agency," and equip managers in health care "to gain an overview of all operations within a hospital, including the management of insurance claims processing, scheduling of testing, equipment needs, and staff assignments."[2]

SAP and Oracle make similar claims, and all three corporations have brought out a host of research documents and "executive briefs" showing how their monitoring and control systems apply to all the principal sectors of the manufacturing and service economies. SAP, for example, has "industry overviews" for higher education, retail, customer relations management, marketing, semiconductors, utilities, manufacturing, banking, and human resource management.[3] CBSs are, then, universal technologies, straddling the boundaries between the public and the private sectors, between manufacturing and services, and between managers, professionals, and front-line workers.

IBM gives a vivid sense of the sheer density of control embedded within its systems. It lists the eight mutually reinforcing "views" of the workplace that managers empowered with its systems can acquire.[4] There is a "scorecard view" that groups together *Key Performance Indicators* (KPIs), such as the sales and profit data for corporate divisions; then "a KPI view" that singles out a particular indicator from this grouping and looks at its performance in greater detail, as at the University of Texas, where the number and value of fee-paying students a professor has attracted to his or her class are monitored and measured.[5] Also a "gauge view" that shows KPIs as "visually emulating the appearance of instruments, like the speedometer in an automobile."

Then a "monitoring view" that shows how well a particular process is being performed in real time and against target; then a "report view" that creates written reports on process performance "relative to a time axis"; then a "dimensional" view that "provides granular details about how especially critical elements of a process are being performed," such as the signing of new clients and the sale of newly introduced products; then an "alert view" that tracks the performance of processes that show signs of going wrong and missing their targets; and finally a "process view" that displays "graphical cues about a user's process statistics."

The key to an understanding of these CBS control systems, and indeed of white-collar industrialism itself, lies buried in the eighth and perhaps most obscure of these "views," the view that displays "graphical cues about a user's process statistics." This eighth view consists of graphical, electronic representations of processes as "an event-driven process chain" in which the mostly computer workstations constituting the process are represented as squares, triangles, or oblongs on the screen, linked to one another in a virtual chain, and so displaying the life cycle of the process from beginning to end. As a symbolic, electronic representation of events in the real world, there is no difference between an electronic "event-process chain" representing a process in manufacturing, such as the movement of a car body along the auto assembly line, and a process chain in the service economy representing the movement of a patient through a hospital or clinic.

Although these electronic chains are potent symbols of the wholesale transfer of industrial methods from the manufacturing to the service economy, there are critical differences between how processes are managed in the two spheres. Counterintuitively, the burden of monitoring and control is much greater in the white-collar economy than that of the blue-collar. Taking the automobile assembly line as the archetypal manufacturing process, the discipline of the line is enforced in the first instance by the repetitive simplicity of the work procedures performed on the line, all meticulously calibrated and timed in advance according to the principles of *Scientific Management.*

This control regime applies irrespective of whether the worker performs a single unvarying routine, as in the early Ford plants, or a routine that varies at the margin, as in Japanese systems of "lean production" now universally adopted in the US auto industry. The moving line itself is also a powerful, all-seeing monitor, because the failure of a worker to perform his assigned task within the designated time immediately shows up in the form of a defective, incomplete workpiece moving on to the next worker on the line.

However, once we move from the blue- to the white-collar line, the iron disciplines of manufacturing fade away, and the human dimension, with all its potential for error and indiscipline, looms larger, and so does the need for a panoptic monitoring regime to pick up on this human waywardness and correct it without delay. Although the electronic "event-process chains" for both the white and the blue lines look the same on management's control "dashboards," the reality each depicts is different. On the white-collar line, the events that populate the process chain are not simply physical movements subject to the full disciplines of Scientific Management.

With the "process events" of customer relations management, human resource management, financial services, and public administration, there is still a place for human judgment and so for human error, along with the human capacity to derail a process and keep it from achieving management's target for its KPIs. With these white-collar processes, there is no physical, mechanical line to ensure that the process events are performed by the right people, in the right order, and in the right amount of time. When the workstations along the line are computers manned by humans, the operator may send the information "workpiece" to the wrong computer, or, if the work divides into subtasks performed on a single computer, the operator may not execute the tasks in the right order and within the designated time frame.

There may even be rogue, unauthorized process "loops" created by employees, so, for example, a "human resource" operative may hire an employee while missing out on steps mandated by the system such as the requirement to "install [the employee] in a learning environment" or "install by special trainer,"[6] or a physician may prescribe a treatment not authorized by an HMO's treatment rule book. These omissions will show up on the electronic process tree, depicting that particular process instance, with the system flagging the process "loops" unauthorized by management.

Although the burden of monitoring and control embedded in

the systems is designed to deal with this human indiscipline, the dominant image of the human that emerges in the texts is the one that the engineers would *like* to be dealing with and so one in which we humans are set alongside the inanimate components of process as abstracted entities fully subject to the manipulation and control of the corporate "process assemblers." These are the engineers who take senior management's preferences for what a process should look like and then come up with the fully elaborated process model. An IBM executive brief for its *Business Activity Monitoring* (*BAM*) software classifies these human and nonhuman components of process as both equally subject to the experimental modeling of the "process assemblers": "When your business analyst is satisfied with the process model, a process assembler can use the graphical tools to pull the services needed from a palette into the process map. The assembler can also drag and drop relationships among data, people, systems and services. The measurement points can be identified and marked."[7]

It is here perhaps that IBM gets us closest to a digital version of Aldous Huxley's *Brave New World* and where, whether we are physicians, fast food workers, middle managers, or Walmart associates, we have become disembodied objects of speed and efficiency joined to these electronic symbols on the screen—symbols that the "process assemblers" then move around as they see fit and with the real, corporeal us having to follow orders like members of a digital chain gang, pushed first one way and then another by our virtual overseers. At the same time, IBM also claims that the system helps managers "perform corrective action based on real time information" when this needs to be done. Corrective actions include "transferring work items" away from workers who may not be meeting their targets and "suspending or terminating the process altogether" so that an investigation of employee error can be undertaken without delay.[8]

Scheer AG of Saarbrücken also pitches its monitoring software for its power to "evaluate various behaviour patterns" and to test whether "secondary paths on the event tree" traversed less frequently by the human agents are "disruptive and should be eliminated."[9] Scheer also draws attention to a new source of disruption—e-mails, chat rooms, discussion forums, blackboards, instance messages, and Web conferences, insisting that the undisciplined use of all these tools has to be replaced with their "order creation in the value creation chain."[10] In the product manual for its Domino Administrator 8, IBM gives a vivid sense of how this "ordered creation" of e-mail use can be achieved. The document has separate sections on topics titled "Tools for Mail Monitoring," "Creating a Mail-Routing Event Generator," "Tracking a Mail Message," "Generating a Mail Usage Report," "Viewing Mail Usage Reports," and "Controlling the Mail Tracking Collector."[11]

The development of such hyperelaborate systems of control testifies to the power of CBS technologies to extend their domain to virtually every human activity performed in the workplace. It also testifies to the temptation, hard to resist, for executives to avail themselves of such powers even if the activities in question may be ill suited to be objects of control. Businesses may need to find ways of picking up on rogue employees who abuse their use of e-mail, but to do so by setting up a panoptic 24/7 system that monitors the entire workforce all the time fosters a culture of mistrust that can only be reciprocated by those who are its objects, and stifles the use of e-mail as a creative outlet for employees otherwise at the mercy of their KPIs.

MEASURES TO CORRECT the unauthorized, the disruptive, and the dysfunctional are the end products of elaborate systems of monitoring and control that invite further exploration of how they work. Again it is the event-driven process chain that is the essential entry point. The most illuminating account of EPCs as control mechanisms is to be found in Scheer AG's volume *Corporate Performance Management,* edited by August-Wilhelm Scheer himself.[12] The account of EPCs in IBM and SAP texts is similar, though less detailed and systematic.

Scheer's monitoring system pivots on the representation of an ideal EPC embedded in the system's memory, the "ideal" being a process archetype that exemplifies exactly how management wants the process to be performed, whether it be the hiring of an employee, the treatment of a patient in a clinic, the packaging of a mortgage, or the assembly of a computer on the line. These archetypal EPCs straddle the frontier between manufacturing and services. They lay down the pathways that the work item, whether physical or virtual, must follow as it wends its ways between the process workstations, the time to be taken for each stage of the process, and the quantitative values for the KPIs that management may attach to the process.

There is no limit to the number of KPIs that management can attach to a process. In the product manual for its *Business Process Management* system, IBM lists thirty-three possible KPIs for a product sales department, including profit margin per transaction, profit per customer, customer average days to pay, contribution to profit by product, and percentage of deliveries on time. Managers may also attach KPIs to a process without the concerned employees knowing about it. IBM's research Red Book for its Websphere Business Monitor V1 describes how a "dashboard KPI" can be created for a manager's "personal use" and is not visible to "other people in the organization."[13]

The core activity of monitoring is an automated comparison between the values of the ideal process instance embedded in the system's memory and the values of the actual performance of employees in the present or the recent past. So, for example, one might compare the time allowed for the processing and final approval of a mortgage application by loan officers set alongside the times actually taken by them. If the comparison is with past process instances, the system can create a composite event tree of these past instances, which will show the average of its KPIs and so also provide evidence that there may be a systemic flaw in how the processes have been executed. Managers can also peer into their subordinates' screens without their knowing it and see how the process is being executed in real "run time." IBM boasts that its monitoring software "enables you to monitor the run-time behaviour of business processes through a web application deployed on IBM WebSphere Process Server, Version 6."[14]

The goal of this monitoring is to find cases where the KPI values attached to real-world "process instances" do not conform to those embedded in the "ideal" process. But the discovery of this malfunction is only the first stage of a three-stage procedure that has to be performed in full if the malfunction is to be properly dealt with. The second stage is the discovery of the causes of error, and the third is the prescription of a cure. The search for causes brings into play a central feature of CBSs, which is their capacity to "*drill down*" to find the causes of dysfunction. Here the manager shifts the electronic images of process instances on his dashboard, each with their own electronic process chain, to find the culprit or culprits.

If the culprit is a single employee, the manager's drill down may end with an examination of the single process instance for which the employee has been responsible. If she has taken too long to perform a "business event," such as the processing of a mortgage application, it may be because she has included procedures not authorized by the rules, and the electronic image of the process instance will show it. If the culprit is a work team or whole department, it may be necessary to create a composite process chain to find exactly where the weak link in the process is to be found. In the CBS product manuals, there is an unrelenting emphasis on the need for speed in the execution of processes and for speed in the detection and correction of process error. In the words of the Gartner Corporation, the leading specialist consultant of the CBS world: "Emerging business activity monitoring and real-time enterprise strategies take the goal of timeliness to its logical conclusion; their aim is instantaneous awareness and appropriate response to events across an entire virtual enterprise."[15]

This preoccupation with speed is simply the transfer to a service-dominated economy of a practice deeply rooted in manufacturing. But while the saving of minutes or seconds on the assembly lines can contribute directly to improved efficiency, there is no such automatic payoff in health care or financial services, where an insistence on speed is just as likely to yield inefficiency in the form of hasty, ill-considered judgments made by professionals harassed by the clock. The logical outcome of this search for speed is the automation of all three monitoring phases, including the prescriptive phase. In the words of Rob Ashe, now general manager of IBM's Business Intelligence and Performance Management Unit: "Any system you want to be fast and pervasive must be automated."[16]

The primary sources reveal that Corporate Panoptics as the monitoring and control of business processes, and Business

Processes Reengineering as their restructuring, have ceased to be distinct activities and are now fused as a continuous activity. In the original reengineering textbooks, the practice comes over as a variant of traditional engineering, where experts descended on the white-collar line, disassembled and reassembled it as if it were made up of real physical objects, and then handed back the improved model to the routine line managers. Now the two management processes have become a single integrated activity in which the inefficiencies revealed by constant monitoring become the raw material for the equally constant activity of reshaping the processes to make them run faster and more efficiently. Although the working lives of employees are deeply affected by this constant change, they have no more say in how it takes place than the hardware and software systems that stand between them and their superiors.

The fusion of Corporate Panoptics and Business Process Reengineering is a leading theme of the primary sources and claimed as a major technological breakthrough for CBSs in the early twenty-first century. So IBM's *Business Dashboard for Managers* first compares "comprehensive actual metrics to established performance measures." The results can then be immediately "fed back into process modelling tools for further analysis and to complete the cycle of continuous process improvement."[17] In Scheer AG's texts, the fusion is described with characteristic detail and thoroughness: "The graphic visualization of actual processes is a patented method for identifying patterns in the execution of processes and pinpointing optimization potential; . . . the running processes are subject to constant improvement through automated measurement and analysis."[18] Again, the human element is completely absent in this perfecting of process.

Managers too are diminished by the coming of Corporate Pan-optics. With the monitoring, analysis, and reshaping of business pro-cesses increasingly automated, the role of middle and lower managers becomes one of gazing at control screens, like shift workers in a highly automated steel mill, and waiting for something unusual to happen. Moreover, their own management performance is as much subject to monitoring as their front-line subordinates, and equally visible to their management superiors. Distinctions between the managers and the managed become increasingly difficult to draw.

Although the heavy automation of new-generation CBSs suggests the workings of artificial intelligence (AI), this is misleading. Every one of the proliferating inventory of rules that govern the work regimes of CBSs is the result of decisions made by senior executives and then embedded in the system by the process assemblers, whether it be the structure of the event-process tree, the timings linking the worksta-tions on the tree, the KPIs attached to the process, or, in the unceasing dialectic between Process Reengineering and Corporate Panoptics, the rules used to assess which process improvements should be autho-rized and which should not. The reverse side of the empowerment of experts is the complete disempowerment of nonexperts.

In this dialectic, and in the equally constant shaping and re-shaping of processes that follow, the nonexperts, whether lower managers or front-line workers, have little or no role to play. They are present on the expert's management dashboards as electronic symbols, not as living entities, and because they are tethered to the symbols, whenever the symbols are "dropped or dragged" onto the dashboard, so are they. They are as much objects of the system as those FedEx packages at a depot shunted onto different conveyer belts according to the codes on their labels.

In the old factory economy, the work routines of mass production applied to the physical movements of workers on the line. But in the contemporary service economy, dominated by CBSs, the targets that matter most are the judgments, human interactions, and even the speech of employees, and the agents of control are these networked computers empowered with workflow and monitoring software, with expert systems attached. Moreover, because employees in such fields as health care, financial services, customer relations, and human resource management are dealing with their fellow humans in all their complexity and contrariness, the rules of the system must proliferate and mutate as they try to cope with the myriad contingencies that can arise in these encounters.

Once these networks are up and running, every aspect of work—the timing of tasks, the sequence in which they are performed, the operation of expert systems—becomes subject to rules that can be altered, elaborated, and enforced at the touch of a managerial keystroke. This applies as much to the work of physicians, middle managers, and the operatives of the Wall Street mortgage machine as it does to the work of Walmart "associates" and call-center agents. What we are witnessing is the emergence of a new white-collar working class, subject to all the regimentation and discipline of its factory predecessor, but lacking the latter's solidarity, its willingness to organize and to fight its cause in the workplace.

2

WALMART AND AMAZON

THESE PRIMARY SOURCES DESCRIBE A SURREAL WORLD OF DIGITAL control in which the human element is in eclipse as just another factor of production. There is a need now to *drill down* and to restore the human by looking at how the systems work in the context of specific industries and workplaces. The wholesale and retail industries are good places to start because both are highly labor intensive and because it is also there that CBSs have achieved some of their most spectacular results, with Walmart leading the way. In its analysis of the growth of US labor productivity between 1995 and 2000—the years of the "new economy" and of the high-tech bubble on Wall Street—the McKinsey Global Institute found that just over half that growth took place in two sectors, wholesale and retail, where Walmart "caused the bulk of the productivity acceleration through ongoing managerial innovation that increased competition intensity and drove the diffusion of best practice."[1]

In the category of "general merchandise," the rise of Walmart took the form of a huge lead in productivity over its competitors, of 44 percent in 1987, 48 percent in 1995, and still 41 percent in 1999, even as competitors began to copy Walmart's methods. In its research McKinsey makes much of Walmart's innovative use of information technology: its early use of computers and scanners to track and replenish inventory; its use of satellite communications to link corporate headquarters in Bentonville, Arkansas, with the nationwide network of Walmart stores; and the labeling of goods with chips, sensors, and Wi-Fi tags to monitor and speed up the movement of goods from factory to warehouse to store.[2] But these technologies are also used to monitor and time the work of employees shifting the goods. In her study *The Quality of Work at WalMart,* Ellen Rosen of the Women's Studies Research Center at Brandeis University describes an especially egregious Walmart work practice that relies on this rigorous monitoring to squeeze the labor budget and keep Walmart wages low.[3]

Each year Walmart provides its store managers with a "preferred budget" for employment, designed to allow managers to staff their stores at adequate levels. But the actual budget imposed on the store managers always falls short of the preferred budget, so that most Walmart stores are permanently understaffed. One store manager explained to Rosen the practical consequences of this: "With the meagre staff he was allowed, it had always been a struggle to keep shelves stocked and the floors shiny, or to get hourly workers to help customers." With each employee having so much work to do, managers assumed that whenever they saw an employee not working, she must be shirking her duties, or "stealing time" from the corporation, a punishable offense.

When the writer Barbara Ehrenreich worked at a Minneapolis Walmart as part of the research for her book on low-wage work, *Nickel and Dimed,* she was told by her boss that "time theft" in the form of associates standing around and talking to one another was his "pet peeve." Later a fellow worker warned Ehrenreich that they could talk only about their work and that anything else counted as "time theft" and was forbidden. Ehrenreich soon found that her boss and his fellow management spies were a constant presence on the shop floor, looking out for time thieves.[4]

Rosen's and Ehrenreich's accounts of the workplace regime at Walmart date from the early 2000's, and since then a burdensome workplace regime for employees has grown even more so as the corporation has tried to compensate for stagnating store sales by cutting costs. This cost cutting has entailed reductions in an already overstretched shop-floor workforce, along with attempts to compensate for these reductions with control technologies that rely on factory disciplines to extract higher output from the remaining workforce. Foremost among these technologies is "Task Manager," a targeting and monitoring system that Walmart began to introduce in its stores from 2010 onward. The system tells employees what to do, how long they have to do it, and whether they have met their target times. Employees sign on to the system by swiping their identity cards on a terminal, as with a credit card, and the system then spits out its instructions.[5]

In a research paper on Walmart's "Productivity Loop," John Marshall of the Capital Stewardships Program of the Union of Food and Commercial Workers (UFCW) found that Task Manager is "an object of scorn among thousands of Walmart associates" who complain that "there is never enough time to complete all the tasks." He cites the case of Maggie, an overnight stocker who had worked in the shoe department of a Walmart store in Southern California for four years: "The Task Manager says you can stock the shoe department in 15 minutes. That's six boxes with 12 pairs of shoes in each box—equals 12.5 seconds per pair of shoes. I'm good at what I do, I know my department very well, but I simply cannot get the work done in the time the systems says." Her colleague Toby, an unloader in the shoe department, added that the system required him to unload one case of shoes per minute; "this is physically impossible."[6]

With Task Manager the process regime at Walmart has begun to fray at the edges, and this deterioration has been noted by stock analysts and the trade press. In February 2011 *Retailing Today* carried a photograph of a Walmart supercenter whose largely empty shelves were reminiscent of the way supermarkets used to look in East Berlin before the Wall came down: virtually empty and this because employees had not had the time to stock the shelves. *Retailing Today* asked: "How could a retailer who so often talks about leveraging its supply chain to deliver upon all sorts of strategic initiatives allow an out of stock situation of such extreme proportions to happen?"[7]

Retailing Today has noted other symptoms of system failure: "unmanned checkout register episodes at multiple Walmarts . . . customers entering a Walmart superstore [who] could find no available shopping carts—since no employees were available to retrieve empty carts from the parking lot." At Walmart's annual shareholders meeting in June 2011, David Strasser of the Philadelphia brokers Janney Montgomery Scott asked Walmart's CEO for the United States, Bill Simon, "Are you still comfortable that you're not taking out too many hours around the register and causing more lines and ultimately losing sales?" Simon acknowledged he was not "comfortable" with the way things were going but, honoring an old Walmart tradition, blamed the store managers: "Some of them try to make their profit number the wrong way. And that's in some of the cases driving some of the action that some people have seen."[8]

One predictable outcome of Walmart's workforce overload has

been an increase in the number of employees who fail to meet the target timings mandated by Task Manager and are then punished. With such a rigged time and motion regime, even the most productive employees can fall short, and when they do Walmart management is ready with an elaborate system of penalties. There are written reprimands in the form of Walmart's own "pink slips"; spoken reprimands in the form of "coachings"; then "decision-making days," or d-days, when an employee must explain why he or she should not be fired; and then finally dismissal itself. The collective impact of these penalties is to keep employees off balance and to undermine their bargaining position when they ask for a pay rise. A former assistant store manager at Walmart reports that with the coming of Task Manager, there had been "a big increase in coachings and in more terminations."[9]

The one-sided politics of the Walmart shop floor are reinforced by a weak code of federal labor laws weakly enforced, and which Walmart has frequently violated in order to push up labor productivity while keeping labor costs down. A chronicle of these violations is to be found in the details of payments Walmart has had to make in penalties to government—federal and local—and awards to plaintiffs in civil suits. Walmart's violations include dismissal of employees for prounion organizing, enforcement of overtime work without pay, denial of paid lunch breaks, and the hiring of illegal immigrants who, once hired, have been confined overnight in Walmart stores. In December 2008 Walmart settled sixty wage class-action suits for $640 million and since then another seven class suits for another $345 million, for a grand total of just under $1 billion.[10]

Walmart has always been a ferociously antiunion company, and the UFCW has yet to organize a Walmart store. Every manager at Walmart is issued with a "Manager's Toolbox to Remaining Union

Free," which warns managers to be on the lookout for signs of union activity, such as "frequent meetings at associates' homes" or "associates who are never seen together . . . talking or associating with one another."[11] The Toolbox provides a special hotline so that managers can get in touch with Walmart's Bentonville headquarters the moment they think employees may be planning to organize a union. A high-powered union-busting team will then be dispatched by corporate jet to the offending store, to be followed by days of compulsory antiunion meetings for all employees.

But not everything has been going management's way in its perpetual conflict with its own workforce. In 2011 OUR Walmart came into existence, with OUR standing for "Organization United for Respect at WalMart." Although strongly supported by the UFCW, OUR Walmart is not a union and is not legally empowered to negotiate with the company on wages and working conditions. It is an association that any Walmart employee can join for a monthly payment of five dollars. Yet under US labor law it has the right to organize walkouts as de facto strikes, and Walmart is prohibited by law from retaliating against the organizers of these actions, although there is strong anecdotal evidence that it does retaliate. The first such strike took place in October 2012 at a Walmart supercenter at Pico Rivera, California, followed in September 2013 by walkouts and protest rallies at Walmart stores in ten US cities, including New York, Chicago, Boston, and Washington, DC.[12]

The willingness of Walmart employees to take what for them is the risky step of joining OUR Walmart and participating in strikes and protest rallies is a warning to Walmart management that there is a limit to how far it can push its workforce without provoking a strong backlash. The results of independent polling carried out in May and June 2011 by Lake Research Partners suggests that Walmart may be getting close to this danger zone. Eighty-four percent of those polled said that they would take a better job if they could find one, three-quarters said that understaffing had undermined customer service, and half said they were living from paycheck to paycheck. The OUR Walmart strikes and rallies also provide a rallying point for civic and political leaders opposing Walmart's further expansion in their communities. The 2012 strike in Southern California had the support of Charles Calderon, majority leader of the California Assembly, and the Reverend Eric Lee, president of the Los Angeles chapter of the Southern Christian Leadership Conference.[13]

The Walmart case provides a spectacular firm-level example of the role of information technology in driving the wages-productivity gap. On the production side, IT supports a global system of logistics that coordinates very efficiently the movements of tens of millions of goods from factories to warehouses and to stores; on the labor side, it promulgates rules that govern the movement and actions of more than a million employees, incorporating the monitoring systems that ensure these actions are performed according to the rules and within the designated time frames. This duality has enabled Walmart to achieve among the highest rates of productivity growth for the entire service economy, while keeping the wages of its "associates" at or barely above the poverty level and while also relying on the taxpayer to keep the children of Walmart employees out of poverty. It is true that this harsh workplace regime yields a payoff in the form of lower prices for Walmart's consumers. But should such convenience for the consumer be purchased with the lost dignity and lost earnings of the Walmart workforce?

WHEN I FIRST did research on Walmart's workplace practices in the early 2000s, I came away convinced that Walmart was the most egregiously ruthless corporation in America. However, ten years later, there is a strong challenger for this dubious distinction—Amazon Corporation. Within the corporate world, Amazon now ranks with Apple as among the United States' most esteemed businesses. Jeff Bezos, Amazon's founder and CEO, came in second in the *Harvard Business Review*'s 2012 world rankings of admired CEOs, and Amazon was third in CNN's 2012 list of the world's most admired companies.[14] Amazon is now a leading global seller not only of books but also of music and movie DVDs, video games, gift cards, cell phones, and magazine subscriptions. Like Walmart itself, Amazon combines state-of-the-art CBSs with human resource practices reminiscent of the nineteenth and early twentieth centuries.

Amazon equals Walmart in the use of monitoring technologies to track the minute-by-minute movements and performance of employees and in settings that go beyond the assembly line to include their movement between loading and unloading docks, between packing and unpacking stations, and to and from the miles of shelving at what Amazon calls its "fulfillment centers"—gigantic warehouses where goods ordered by Amazon's online customers are sent by manufacturers and wholesalers, there to be shelved, packaged, and sent out again to the Amazon customer.

Amazon's shop-floor processes are an extreme variant of Taylorism that Frederick Winslow Taylor himself, a near century after his death, would have no trouble recognizing. With this twenty-first-century Taylorism, management experts, *scientific managers,* take the basic workplace tasks at Amazon, such as the movement, shelving, and packaging of goods, and break down these tasks into their subtasks, usually measured in seconds; then rely on time and motion studies to find the fastest way to perform each subtask; and then reassemble the subtasks and make this "one best way" the process that employees must follow.

Amazon is also a truly global corporation in a way that Walmart has never been, and this globalism provides insights into how Amazon responds to workplaces beyond the United States that can follow different rules. In the past three years, the harsh side of Amazon has come to light in the United Kingdom and Germany as well as the United States, and Amazon's contrasting conduct in America and Britain, on one side, and in Germany, on the other, reveals how the political economy of Germany is employee friendly in a way that those of the other two countries no longer are.

Amazon, like General Electric and Walmart, prides itself as a self-consciously ideological corporation, with Jeff Bezos and his senior executives proclaiming an "Amazon Way" that can illuminate the path forward for less innovative businesses. In December 2009 Mark Onetto, chief of operations and customer relations at Amazon and a close collaborator of Bezos, gave an hourlong lecture on the Amazon Way to master's of business administration students at the University of Virginia's Darden School of Business.[15] Onetto is a disconcerting figure, because once he starts talking, style and substance are in sharp contrast. He is French born, and he still speaks with the rather faded insouciance of Maurice Chevalier and "Gay Paree," and he makes much of this in his lecture. But there was nothing gay (in the traditional sense) or insouciant about the Amazon workplace that Onetto described for UVA's MBA candidates.

Like most such corporate mission statements, Onetto's uses a coded language that hides the harshness of his underlying message, which needs translation along with a hefty reality check. As with Walmart so at Amazon, there is a quasi-religious cult of the customer as an object of "trust" and "care"; Amazon "cares about the customer," and "everything is driven" for him or her. Early in the lecture, Onetto quotes Bezos himself as saying, "I am not selling stuff. I am facilitating for my customers to buy what they need."[16]

Amazon's larding of its customer cult with the moral language of "care" and "trust" comes with a strong dose of humbug because Amazon's customers are principally valued by the corporation as mainstays of the bottom line, and not as vehicles for the fulfillment of personal relationships. There is still more humbug in the air because Amazon treats a second significant grouping of men and women with whom it has dealings—its employees—with the very opposite of care and trust. Amazon's employees are almost completely absent from Onetto's lecture, and they make their one major appearance when they too are wheeled in as devotees of the cult of the customer: "We make sure that every associate at Amazon is really a customercentric person, that cares about the customer."[17]

But as so often in Amazon's recent history, it has been in Germany that this humbug has been stripped away and the true role of the "cult of the customer" has become clear. In its US and UK fulfillment centers, Amazon management is hegemonic. There is no independent employee voice to contest management's demands for increased output unmatched by increases in real wages. But in Germany Amazon has to deal with work councils (*Betriebsrat*); a powerful union, the United Services Union (Vereinte Dienstleistungsgewerkschaft, or Ver.Di), with 2.2 million members; and high officials of the federal and state governments more closely aligned with labor than their counterparts in the United States and the United Kingdom.

When in December 2012 the Ver.Di representatives in Leipzig called on the management of Amazon's local center to open negotiations on wage rates and an improvement of working conditions, and especially for temporary workers who are badly exploited at Amazon, management refused on the grounds that employees should be "thinking about their customers" and not about their own selfish interests.[18] This was treated with derision on the union side, but at all Amazon's centers, and especially those in the United States and the United Kingdom, the cult of the customer is a serious matter and provides the rationale for the extreme variant of scientific management whose purpose, as at Walmart, is to keep pushing up employee productivity while keeping hourly wages at or near poverty levels.

As at Walmart, Amazon achieves this with a regime of workplace pressure, in which targets for the unpacking, movement, and repackaging of goods are relentlessly increased to levels where employees have to struggle to meet their targets and where older and less dextrous employees will begin to fail. As at Walmart, there is a pervasive "three strikes and you're out" culture, and when these marginal employees acquire too many demerits ("points"), they are fired.

Amazon's system of employee monitoring is the most oppressive I have ever come across and combines state-of-the-art surveillance technology with the system of "functional foreman," introduced by Taylor in the workshops of the Pennsylvania machine-tool industry in the 1890s. In a fine piece of investigative reporting for the London *Financial Times,* economics correspondent Sarah O'Connor describes how, at Amazon's center at Rugeley, England, Amazon tags its employees with personal sat-nav (satellite navigation) computers that tell them the route they must travel to shelve consignments of goods, but also set target times for their warehouse journeys and then measure whether targets are met.[19]

All this information is available to management in real time, and if an employee is behind schedule she will receive a text message pointing this out and telling her to reach her targets or suffer the consequences. At Amazon's depot in Allentown, Pennsylvania (of which more later), Kate Salasky worked shifts of up to eleven hours a day, mostly spent walking the length and breadth of the warehouse. In March 2011 she received a warning message from her manager, saying that she had been found unproductive during several *minutes* of her shift, and she was eventually fired.[20] This employee tagging is now in operation at Amazon centers worldwide.

Whereas some Amazon employees are in constant motion across the floors of its enormous centers—the biggest, in Arizona, is the size of twenty-eight football fields—others work on assembly lines packing goods for shipping. An anonymous German student who worked as a temporary packer at Amazon's depot in Augsburg, southern Germany, has given a revealing account of work on the line at Amazon. Her account appeared in the daily *Frankfurter Allgemeine Zeitung,* the stern upholder of German financial orthodoxy and not a publication usually given to accounts of workplace abuse by large and powerful corporations.[21] There were six packing lines at Amazon's Augsburg center, each with two conveyor belts feeding tables where the packers stood and did the packing. The first conveyor belt fed the table with goods stored in boxes, and the second carried the goods away in sealed packages ready for distribution by UPS, FedEx, and their German counterparts.

Machines measured whether the packers were meeting their targets for output per hour and whether the finished packages met their targets for weight and so had been packed "the one best way." But alongside these digital controls there was a team of Taylor's "functional foremen," overseers in the full nineteenth-century sense of the term, watching the employees every second to ensure that there was no "time theft," in the language of Walmart. On the packing lines there were six such foremen, one known in Amazonspeak as a "co-worker" and above him five "leads," whose collective task was to make sure that the line kept moving. Workers would be reprimanded for speaking to one another or for pausing to catch their breath (*Verschnaufpause*) after an especially tough packing job.[22]

The functional foreman would record how often the packers went to the bathroom and, if they had not gone to the bathroom nearest the line, why not. The student packer also noticed how, in the manner of Jeremy Bentham's nineteenth-century panopticon, the architecture of the depot was geared to make surveillance easier, with a bridge positioned at the end of the workstation where an overseer could stand and look down on his wards.[23] However, the task of the depot managers and supervisors was not simply to fight time theft and keep the line moving but also to find ways of making it move still faster. Sometimes this was done using the classic methods of Scientific Management, but at other times higher targets for output were simply proclaimed by management, in the manner of the Soviet workplace during the Stalin era.

Onetto in his lecture describes in detail how Amazon's present-day scientific managers go about achieving speedup. They observe the line, create a detailed "process map" of its workings, and then return to the line to look for evidence of waste, or *Muda*, in the language of the Toyota system. They then draw up a new process map, along with a new and faster "time and motion" regime for the employees. Amazon even brings in veterans of lean production from Toyota itself, whom Onetto describes with some relish as "insultants," not consultants: "They are really not nice. . . . [T]hey're samurais, the real last samurais, the guys from the Toyota plants."[24] But as often as not, higher output targets are declared by Amazon management without explanation or warning, and employees who cannot make the cut are fired. At Amazon's Allentown depot, Mark Zweifel, twenty-two, worked on the receiving line, "unloading inventory boxes, scanning bar codes and loading products into totes." After working six months at Amazon, he was told, without warning or explanation, that his target rates for packages had doubled from 250 units per hour to 500.[25]

Zweifel was able to make the pace, but he saw older workers who could not and were "getting written up a lot" and most of whom were fired. A temporary employee at the same warehouse, in his fifties, worked ten hours a day as a picker, taking items from bins and delivering them to the shelves. He would walk thirteen to fifteen miles daily. He was told he had to pick 1,200 items in a ten-hour shift, or 1 item every thirty seconds. He had to get down on his hands and knees 250 to 300 times a day to do this. He got written up for not working fast enough, and when he was fired only three of the one hundred temporary workers hired with him had survived.[26]

At the Allentown warehouse, Stephen Dallal, also a "picker," found that his output targets increased the longer he worked at the warehouse, doubling after six months. "It started with 75 pieces an hour, then 100 pieces an hour. Then 150 pieces an hour. They just got faster and faster." He too was written up for not meeting his targets and was fired.[27] At the Seattle warehouse where the writer Vanessa Veselka worked as an underground union organizer, an American Stakhnovism pervaded the depot. When she was on the line as a packer and her output slipped, the "lead" was on to her with "I need more from you today. We're trying to hit 14,000 over these next few hours."[28]

Beyond this poisonous mixture of Taylorism and Stakhnovism, laced with twenty-first-century IT, there is, in Amazon's treatment of its employees, a pervasive culture of meanness and mistrust that sits ill with its moralizing about care and trust—for customers, but not for the employees. So, for example, the company forces its employees to go through scanning checkpoints when both entering and leaving the depots, to guard against theft, and sets up checkpoints *within* the depot, which employees must stand in line to clear before entering the cafeteria, leading to what Amazon's German employees call *Pausenklau* (break theft), shrinking the employee's lunch break from thirty to twenty minutes, when they barely have time to eat their meal.[29]

Other examples include providing UK employees with cheap, ill-fitting boots that gave them blisters;[30] relying on employment agencies to hire temporary workers whom Amazon can pay less, avoid paying them benefits, and fire them virtually at will; and, in a notorious case, relying on a security firm with alleged neo-Nazi connections that, hired by an employment agency working for Amazon, intimidated temporary workers lodged in a company dormitory near Amazon's depot at Bad Hersfeld, Germany, with guards entering their rooms without permission at all times of the day and night. These practices were exposed in a television documentary shown on the German channel ARD in February 2013.[31]

Perhaps the biggest scandal in Amazon's recent history took place at its Allentown, Pennsylvania, center during the summer of 2011. The scandal was the subject of a prizewinning series in the Allentown newspaper, the *Morning Call,* by its reporter Spencer Soper. The series revealed the lengths Amazon was prepared to go to keep costs down and output high and yielded a singular image of Amazon's ruthlessness—ambulances stationed on hot days at the Amazon center to take employees suffering from heat stroke to the hospital. Despite the summer weather, there was no air-conditioning in the depot, and Amazon refused to let fresh air circulate by opening loading doors at either end of the depot—for fear of theft.

Inside the plant there was no slackening of the pace, even as temperatures rose to more than 100 degrees.[32]

On June 2, 2011, a warehouse employee contacted the US Occupational Safety and Health Administration to report that the heat index had reached 102 degrees in the warehouse and that fifteen workers had collapsed. On June 10 OSHA received a message on its complaints hotline from an emergency room doctor at the Lehigh Valley Hospital: "I'd like to report an unsafe environment with an Amazon facility in Fogelsville. . . . Several patients have come in the last couple of days with heat related injuries."

On July 25, with temperatures in the depot reaching 110 degrees, a security guard reported to OSHA that Amazon was refusing to open garage doors to help air circulate and that he had seen two pregnant women taken to a nursing station. Calls to the local ambulance service became so frequent that for five hot days in June and July, ambulances and paramedics were stationed all day at the depot. Commenting on these developments, Vickie Mortimer, general manager of the warehouse, insisted that "the safety and welfare of our employees is our number-one priority at Amazon, and as general manager I take that responsibility seriously." To this end, "Amazon brought 2,000 cooling bandannas which were given to every employee, and those in the dock/trailer yard received cooling vests." [33]

WITH WALMART'S AND Amazon's business model, the workplace practices that raise employee productivity to very high levels also keep employees off balance and thus ill placed to secure wage increases that match their increased output. The "cult of the customer" preached by both corporations is a scented smoke screen thrown up to hide this fact. Apart from the model's intensive use of IT, there is not much to distinguish its methods from those of the primitive American and European capitalism of the late nineteenth and early twentieth centuries. On both sides of the Atlantic, these excesses were harbingers of the rise of the labor movement and the political Left, both revolutionary and democratic, with the movements strongly focused on relations between capital and labor as the central issue of politics and society.

In the United States and the United Kingdom, the parties of the center Left, the Democrats and the Labour Party, have today lost this focus, and the labor movements in both countries are in long-term decline. But in Germany the labor movement remains strong, and on workplace issues the mainstream political parties, the Christian Democrats as well as the Social Democrats, are well to the left of their American and British counterparts. This became apparent following the scandal at Amazon's Bad Hersfeld depot in 2012, when security guards allegedly forced their way into dormitories housing temporary Amazon employees and intimidated them. Amazon faced what *Der Spiegel* called a *Shitstorm* and was strongly criticized by the federal minister of labor, the prime minister of the state government of Hesse, the head of the Labor Office in Hesse, as well as the Social Democratic Party opposition in the federal and state parliaments.[34]

Amazon was on the defensive, and in an interview with *Spiegel Online* that followed the scandal, Amazon's local CEO, Ralf Kleber, distanced himself from the managerial absolutism of Bezos and Onetto in saying that he would welcome the setting up of more work councils (*Betriebsrat*) at Amazon depots.[35] The services union, Ver.Di, was also a beneficiary of the Amazon *Shitstorm*. The union's goal is to organize the whole Amazon workforce in Germany, negotiate wage increases with Amazon management, improve the working conditions of temporary employees, and blunt Amazon's more oppressive workplace practices. In a German political and social context, it has a good chance of succeeding. Such success would, however, raise issues of ethics and economics that apply equally in a US and UK setting.

Union success would unquestionably raise Amazon's costs and slow the growth of employee productivity. Wages would begin increasing in line with employee productivity, and productivity growth itself would slow as the union and the *Betriebsrat* together blunted Amazon's practice of pushing employees to the limit and beyond. We can be sure that at this point, Amazon would play the "cult of the customer" for all its worth and would do the same in an American setting if faced with the same challenge. So customers would have to start paying more for their packages and could no longer be absolutely certain of receiving delivery of them the very next day.

But should these marginal benefits to customers really be purchased at the price of a system that treats employees as untrustworthy human robots and relies on intimidation to push them to the limit, while denying them the rewards of their own increased efficiency? This is not a choice to be made solely with the economist's narrow calculations of monetary costs and benefits. In quantitative, monetary terms, the cost to Amazon customers of a benign reengineering of the company would far outweigh the monetary benefits to employees. But what is the real value of such customer inconvenience when set alongside the value lost with the millions of lives damaged by Walmart, Amazon, and their ilk?

3

A FUTURE
FOR THE MIDDLE CLASS?

THESE AMAZON AND WALMART HISTORIES ARE PRIME EXAMPLES of how in the early twenty-first century, state-of-the-art information technologies can be used to re-create the harsh, driven capitalism of the pre–New Deal era. With their reliance on tens of thousands of workers to shift goods in stores and warehouses, the two corporations depend heavily on a steady supply of unskilled labor very much in the manner of early-twentieth-century industrial sweatshops. But in their capacity to track employee performance, to speed it up, to measure it against targets, managers at Walmart and Amazon are empowered in ways that their predecessors of a century ago could only dream of.

The countervailing powers of labor unions are needed to offset this management hegemony and to defend the dignity of employees—in securing wage increases that match their increased output, in placing limits on shop-floor "speedup," and in protecting employees against unfair and arbitrary dismissal. The need for a strong union presence is all the greater because most work in Walmart stores and Amazon "fulfillment centers" is by its very nature unskilled, and employees cannot realistically expect that technology will at some time in the future enhance their skills and add to their earning power.

But there are also sectors of the manufacturing and industrialized service economies where the relationship between technology and skill is more complex, where a given technology can coexist with different levels of skill, and where the relationship between technology and skill that prevails reflects a particular business history and culture. In the American case the close identification of information technology and of CBSs with an authoritarian, mass-production model, heavily influenced by Taylorist Scientific Management, is an outcome of this history and culture and is not an intrinsic quality of the technologies themselves. Again, as with Amazon, it is the contrast between American and northern European and specifically German workplace practices that can show that this American model is not a universal norm but an American phenomenon whose reach is finite and can be contested.

Perhaps the best way to illustrate this pluralism is to compare production regimes from the two industrial cultures. They may rely on similar technologies to turn out broadly similar goods, but the histories embedded in the two models make for very different relationships between men and machines, with these differences showing up in the differing sophistication and quality of goods produced under the two regimes. At play here is the distinction between craft labor, which is skilled, and industrial labor, which is not skilled, a distinction that has its origin in the machine age but continues to yield valuable insights in the digital present.

In the fall of 1992, two years after the reunification of Germany,

I visited the former East German city of Chemnitz, known as Karl-Marx-Stadt during the years of the German Democratic Republic (GDR). In prewar, united Germany, Chemnitz had been a center of the German machine-tool industry, but following World War II and the division of Germany, much of the industry had migrated westward to the Rhineland *Land* of Baden-Württemberg. But a residual East German machine-tool industry had remained intact in Chemnitz, and I was interested in seeing how it was getting along following German reunification in October 1990. I visited a company that was considered one of the best of the East German industry and at the time was under the control of the Treuhandanstalt, the state holding company that after reunification had been given the task of taking control of those East German companies that, in the view of West German experts, could be turned around and eventually sold off to a West German or foreign buyer.

I had originally become interested in the West German engineering industries, and the machine-tool industry in particular, when in the 1970s and early 1980s they invaded the British market and, with their superior quality and performance, effectively destroyed the old British engineering industries that had their origins in the Industrial Revolution and had managed to survive World War II more or less in tact. A remarkable series of research papers published by the London-based National Institute of Economic and Social Research showed in detail how the superior skills and more disciplined work habits of the German engineering workforce enabled the German industries to beat the British on product reliability and delivery times and to market engineering products of a technological sophistication that the less skilled British workforce could not match.[1]

My most memorable encounter with the German machine-tool industry took place in the spring of 1989 when, shortly before the coming down of the Berlin Wall, I was in the press party that accompanied Mikhail Gorbachev to Stuttgart, capital city of Baden-Württemberg and so at the heart of the West German machine-tool industry. Because the supposed achievements of the Soviet machine-tool industry had played a central role in the mythology of Soviet industrialization, the West German machine-tool industry had organized a mini exhibition at Stuttgart University, where the industry's most advanced products were on display so that Gorbachev could inspect them and perhaps be persuaded to have his Machine Building Ministry back in Moscow place orders for some of them.

These were the world's most advanced grinding machines and special-purpose machines for making very high-precision components for machining systems and for machine tools themselves. The machines were objects of elegance, even beauty, and I watched carefully as Gorbachev inspected them. He was used to huge, clunky Soviet machines so heavy that they sometimes fell through factory floors that could not bear their weight. Gorbachev was deeply impressed, even mesmerized by the German machines, and I wondered what was going through his mind; was this the moment when he realized that the backwardness of Soviet industry was beyond remedy, or, in the traditional Soviet manner, did he think that with one big push led by himself, the gap could be closed?

Along with the beauty of the machines and Gorbachev's interest in them, another remarkable feature of the Stuttgart event was that the task of explaining to Gorbachev how the machines worked and what they did was entrusted not to executives of the parent machine-tool companies, but to the craft workers, the master machinists, or *Meisters*, who had actually built the machines. Teams of them clustered around each machine and took their turn explaining its finer points to Gorbachev. These machining *Meisters* were at the pinnacle of the German craft-worker hierarchy and not just because of the preeminent role of the engineering industries in the German export economy. They were graduates of a highly demanding system of study and on-the-job training that can take as long as ten years to complete.

This scene was very much on my mind when I visited the Treuhand-owned machine-tool company in Chemnitz in the fall of 1992. I expected the company still to be scarred by the heavy hand of the East German industrial and planning ministries whose rule the company had endured for forty years. But I was wrong, and this was in part because the East German regime had hung on to the prewar system of German industrial training from which the West German system itself was descended. During the GDR years, the caliber of the training had inevitably lagged behind its West German counterpart, but it still provided a base of reliable expertise that the Treuhand management brought in from West Germany had been able to work with in upgrading the skill levels of the Chemnitz workforce toward West German levels.

This history is a necessary background for an understanding of what I came across on the shop floor at the Chemnitz plant, which was as remarkable in its way as the event with Gorbachev at Stuttgart. I saw a group of machinists clustered around a pile of blueprints. The machinists explained that the arrival of the blueprints at the plant that very morning was a major event in the post-GDR history of the company. The blueprints had come from Daimler-Benz in Stuttgart, and they contained the designs for engine components for a new Mercedes-Benz S-class sedan. But the arrival of the blueprints did not mean that the company had already won an order from Mercedes, a very significant achievement for a Chemnitz engineering company.

The blueprints were there as part of a test to see whether the standards of machining at the plant were advanced enough for Mercedes-Benz to take the risk of placing an order with a former East German company. It was the task of the Chemnitz team to work out how best to machine the Mercedes components, mindful always of the need to achieve very high accuracies without undue sacrifice of machining speed. Working with component prototypes, they had to decide how best to allocate the machining task between different computer-controlled machine tools and work out the machining program for each of them, which would then be fed into the machine's digital controller. They also had to know how to alter their own programs in light of orders for component design changes that would be coming down from Mercedes-Benz during the lifetime of the S-class model.

There was no one from management supervising their work. Already the West German workplace model of strong unions (in this case IG Metall), work councils, and employee-management co-determination was making itself felt in operation at the Chemnitz plant and shaping its day-to-day division of labor. In the case of the Mercedes components, it was ultimately up to the Chemnitz management to decide whether the component prototypes were good enough to send to Mercedes for approval. But it was for the skilled machinists to create the components in the first place. I later found out that the Chemnitz machinists were successful on both counts.

Let's now fast-forward and move from the former Karl-Marx-Stadt to two American industrial communities with absolutely no connection to Marx: Waterloo, Iowa, home to a John Deere plant making agricultural equipment, and Peoria, Illinois, home of the giant Caterpillar plant, making heavy equipment for the construction industry, such as bulldozers, excavators, and heavy trucks. My interest in visiting the two midwestern plants in 2001 was to look at the role of labor in two large mass-production plants already deploying the manufacturing versions of CBSs. But with my Chemnitz experience in mind, I was also interested in seeing how the skilled work of component prototyping was done in an American industrial setting.

It was done very differently. Right at the outset of my visit to the John Deere plant, I was introduced to a gentleman called Merrill Oakman, who described himself as "vice president for machining." It was he who drew up the machining programs for component prototypes and who, for the actual prototype machining, relied on a team of machinists who were under his direct control and were not part of the shop-floor machining workforce. The same arrangement existed at Caterpillar's Peoria plant, with the difference that the prototyping machinists were located at a facility well away from the main manufacturing plant, so that there could be no day-to-day interaction between these elite machinists, who of necessity had to be skilled, and the main shop-floor workforce, which was kept largely unskilled.

At the Peoria plant these arrangements reflected the strong antiunion bias of Caterpillar management and its poor relations with a workforce still organized by the United Automobile Workers union. At UAW Local 974 in Peoria, I met a dozen UAW members who, according to the union, had been unfairly dismissed by Caterpillar and whose cases were pending at the National Labor Relations Board, an interminable process. Both the Caterpillar and the Treuhand models relied on identical technologies to produce the prototype components. But in the Chemnitz model, power and responsibility lay principally with the skilled shop-floor workforce, with management signing off on their work. In the Caterpillar model, the shop-floor workforce was excluded from the prototyping work, with power and skill vested in management, assisted by small teams of machinists under tight management control. The moral of these histories is that although the rootedness of most American CBS regimes in mass production may seem immutable, it is not.

THAT SIMILAR ENGINEERING technologies can accommodate differing regimes of workplace skills has been evident since the days when the rise of Fordism displaced the craft workshops of the earliest automakers. That the technologies of *Customer Relations Management* can accommodate the same workplace duality is less obvious, since call centers as the focus of the CRM world are not usually thought of as places where skills are in high demand. But in the US service economy, CRM technologies can also have complex relations to labor and to skill. Here too technologies can go either way in their workplace impact. Again, the best way to show this is by describing contrasting work regimes that can display a plurality of workplace values.

The case history that follows is drawn from my own working life. It describes an encounter that lasted just twenty-five minutes, yet this tiny molecule in the vast ocean of CRM already contains the DNA of multiple technologies, their variety of possible uses, and how these different uses can require very different levels of skill from the employees concerned. In this history the Japanese corporation Toshiba is the agent of Scientific Management and Microsoft the agent of skill, proof that all is not yet lost in the US service economy.

Most readers will, I suspect, have had a long and mostly negative experience of CRM and its 1-800 numbers. One summer day in 2004, my laptop computer began to exhibit alarming symptoms of digital illness: was the disk drive about the die, was the operating system—Windows XP—under attack from viruses or hackers on the 'Net, or had the computer simply come to the end of its natural life? To find a diagnosis and remedy, I contacted a Microsoft call center that turned out to be located in the Maritime Provinces of Canada. After a discussion lasting some fifteen minutes, the Microsoft CRM agents—by this time there were two of them—decided that the cause of breakdown might be on the hardware side and suggested that we bring in an agent from Toshiba as the laptop's manufacturer, which we did.

There then followed a three-sided dialogue that provided an unusual and illuminating example of how two leading IT corporations can rely on the same combination of CRM technologies yet how each could deploy them in ways that made very different demands on the CRM agents involved. As owner of the diseased computer, I myself was a participant in the CRM processes because I had to provide an intelligible account of the sick computer's symptoms that the agents could then work with. I was also a privileged observer of the processes because I also happened to be a scholar of CRM, someone who could draw on past visits to call centers and on years of immersion in the in-house literature of the industry, such as the e-magazine *Call Center Solutions,* as well as the texts of the contrarian critiques of corporate CRM, such as the work of Erik Vinkhuzen and Jack Whalen at Xerox PARC.[2]

On this occasion there were five technologies integral to the CRM processes and available to both the Microsoft and the Toshiba agents. There was, of course, my computer itself, whose sickness both corporate teams set out to diagnose and cure; there were the corporate databases for both the hardware and the software involved, including case histories of how they both could go wrong and how they could be restored; there was the monitoring software that specified target times for the CRM transactions and with the capacity to measure performance against target; there were also *Management Information Systems (MIS),* which measured agents' performance over time and could integrate this data with those for other employees within the work team or department; and finally there were the agents' "people skills," which, though not strictly speaking a technology, were nonetheless bound up with discussion of the technologies and indispensable for effective CRM.

From the very outset of the Microsoft segment of my CRM encounter, it became apparent that the corporation's CRM regime differed very significantly from those I had become used to over the past ten years. After I had given my hesitant and no doubt incoherent account of the computer's problems, the Microsoft agent—then on his own—was empowered to say: "This is a serious problem, and we're going to have to give it as much time as we need." Eureka! Neither before nor since can I recall a CRM agent setting aside the time factor with such dispatch, and this simple announcement had a transformative effect on everything that followed. It meant that the corporation's monitoring technologies focused on the outcome of its CRM transactions and not on the amount of time it took to achieve them, thus empowering the agents to exercise their skills free of the tyranny of the ticking clock.

Once the dialogue between myself and the Microsoft agents had gotten under way, it also became very clear that their knowledge of the Windows XP system, including the ways in which it could go wrong, was encylopedic and authoritative and that the online database available to them was a backup system that they referred to only when needed. Finally, there were the agents' people skills, and these were not extrinsic to their technical skills, but an integral part of them. Tact was needed as they walked me through my fragmented history of computer breakdown and as they searched for key items of evidence that I had overlooked, patience as they asked me to perform simple tests on the computer that I had a hard time getting right the first time, and also even empathy as they reassured me that the computer could, like some ailing relative, be saved.

This CRM idyll came to an abrupt end when the Microsoft agents concluded that there was a hardware dimension to the problem and that Toshiba had to be brought into the picture. Once the Toshiba agent came onto the line, he brought with him the very different ethos of a CRM regime subject to rigorous industrial disciplines. The critical difference between the Microsoft and Toshiba CRM regimes was that the online database available to the Toshiba agent was not a passive backup for the agent's own expertise, which he might call upon as he needed. Rather, it was a *substitute* for his expertise, an automated and authoritative expert system embodying rules drawn up by expert scientific managers that governed every aspect of the CRM transaction.

The sure sign of this, familiar to me from countless such encounters, was that, once the Toshiba agent had established to his own satisfaction what the symptoms of hardware malfunction were, he embarked upon a line of questioning with wording that was clearly not his own and with a sequencing of questions dictated by the rules of the expert systems and not by the answers I gave to his successive questions. Time was also at a premium, as question and answer followed one another in quick succession and both agent and customer (myself) felt the panoptic presence of Toshiba's monitoring systems with its target timings and with penalties for the agent if he failed to meet them.

The presence and eventual intervention of the Microsoft agents in the dialogue were problematic for the Toshiba agent, both because it disturbed the line of questioning dictated by the Toshiba expert system and because, freed of time constraints, the Microsoft agents wanted to pursue their own line of questioning about the relationship between the hardware and software aspects of the problem with a care and detail that Toshiba's assembly-line regime could not match.

My instinctive reaction at this point was to be irritated with the Toshiba agent for messing up the CRM processes with his clumsy interventions. But this was unfair. He was a subject of the system and of Scientific Management, someone denied the opportunity to make something of his job. He was "skilled" and "empowered" only in the distorted language of reengineering. But the presence of the Microsoft agents on the line was proof that this diminished status was a consequence of management choice, not an inevitable outgrowth of the CRM technologies.

The moral of these histories is that although the rootedness of American CBS regimes in the practices of mass production and Scientific Management may see immutable, it is not. There are alternatives, and the future of much of the remaining American middle class depends on whether American business embraces these alternatives. The middle class in question has been defined by President Obama and others as the upper reaches of the old industrial working class that belonged to unions and earned good wages and benefits.[3] The three postwar decades ending in the mid-1970s were the halcyon days of this blue-collar middle class. These fortunate workers—by today's standards—could move to the suburbs, own their own homes, drive their own cars, even send their children to college.

Two of the few remaining strongholds of this blue-collar middle class are the aerospace and defense industries, where lavish orders from the Department of Defense have bolstered corporations like Boeing and Lockheed Martin, at least through the early 2000s. But elsewhere in US manufacturing, the numbers tell a story of decline, with total US employment in manufacturing falling from 18.7 million in 1980 to 11.9 million in 2012.[4] Although the call for a middle-class restoration pervades the rhetoric of political Washington, particularly on the Democratic side, a resigned fatalism on the issue is never far from the surface, linked to globalization and to the IT revolution as lethal enemies of a revived American middle class: globalization with its transfer of production and jobs to the lower-wage developing world, led by China; the IT revolution with its automation and its displacement of humans by smart machines. But the argument of this chapter suggests that this fatalism is overdone. Globalization forces the advanced economies to migrate in their industrial profiles to levels that are beyond the reach of China and the developing world, and information technology gives them the means to do this.

In a range of IT industries, the United States has achieved this upward mobility: Apple in consumer electronics, Google and Facebook in the software of the Internet, and IBM and Oracle in the technologies of CBSs, as we have seen. But these industries, notoriously, have not been creating a revived version of the industrial middle class on American soil. Apple retains an elite core of managers, designers, and technicians in the United States, but outsources most of its manufacturing to China and Southeast Asia. Google and Facebook can dispatch with manufacturing altogether. From here it is a small step to conclude that the American middle class in its industrial incarnation is indeed doomed never to return and that

President Obama would do well to admit this and stop engendering false hopes with his rhetoric of a middle-class revival.

But this pessimism leaves out a sprawling industrial world that is technologically advanced, skill intensive for the industrial middle class, and guaranteed a central role in the emerging global economy. This is the world dominated globally by the German engineering industries and was on show when I visited the Stuttgart University with Gorbachev and, in embryonic form, at the Treuhand machine-tool plant in Chemnitz. This is the world of high-end machine tools, special-purpose machinery, machining systems, optical and measuring equipment, high-end ships, high-speed trains, and high-end automobiles—BMWs and Mercedes-Benzes.

Common to all these products is a high degree of customization that the methods of mass production cannot accommodate and require a big input of skilled labor at all levels of the enterprise. Strong future demand for these mostly investment goods is virtually guaranteed because they are needed for the industrial infrastructure that China and other developing economies must create in order to achieve their own industrialization. Success in these industries helps explain why, despite the Great Recession, German exports of machinery and electrical equipment increased from $252 billion in 2005 to $359 billion in 2011.[5] These sectors support the existence of an industrial middle class, and some of its members were present that day at Stuttgart University.

The continued existence of its defense and aerospace corporations is proof that the United States can succeed in these sectors; the challenge is to extend this success to markets where the cushion of government contracts is absent. Standing in the way of such an expansion is the withering away in the United States of the vocational infrastructure—technical schools, apprenticeships—needed to supply skilled labor in sufficient quantities. When in the early 2000s I visited the Swiss-German company Hydromat, makers of highly customized machining systems at their US plant in St. Louis, I was told by managers that they had searched the US labor markets in vain for skilled machinists.

In the end they had to train their own machinists, contracting with a local community college for the academic side of the training. President Obama has singled out community colleges as providers of advanced industrial skills for US industry, not a role they were originally designed for.[6] But community colleges are mainly funded by state governments, and their budgets have been severely cut during the years of the Great Recession.[7] With the fiscal deadlock in Washington, it is far from clear that President Obama or his successor can come up with the funds to turn community colleges into technical schools. But the middle-class revival depends on it.

My encounter with the Microsoft agents in the Canadian Maritimes shows that call centers too can be centers of skill and that it is within the power of businesses to make them so. The fact that most businesses choose not to is for them a matter of deliberate choice, shaped by their drive to keep labor costs low and to fill the ensuing skills gap with the databases, digital scripts, and expert systems embedded within CBSs. Our daily struggles with the 1-800 world are proof that this approach does not work, that it wastes time, demeans employees, and is a source of endless frustration for the customer.

The industry is in urgent need of renewal because it is, despite its present sorry state, an emblematic industry of the digital age. It is fast growing, intensive in its use of IT, and, with its huge workforce, an important employer for Americans who have not been to college. With the decline of blue-collar, middle-class unionized factory workers, there is a need to create good jobs for the non–college educated, jobs that are skilled, pay well, and offer the prospects of a career. High-performance call centers can provide such opportunities. But first the industry has to rid itself of the industrial legacy of Frederick Winslow Taylor and Scientific Management.

4

MANAGING THE HUMAN RESOURCE

WHATEVER STRESSES CBSs MAY INFLICT UPON THE WORKFORCES subject to their disciplines—and in the case of Walmart and Amazon, the stress is considerable—CBSs have a natural habitat within which they can often be relied upon to yield strong productivity growth. But this habitat is enclosed within a frontier beyond which the use of CBSs becomes highly problematic, to the point where productivity increases are hard to come by, or are achieved only at a price far outweighing whatever is gained on the productivity side. This frontier does not conform to the economists' conventional distinction between the manufacturing and service economies, but rather conforms to the distinction between those sectors of the economy that are engaged in the manufacture, distribution, and sale of goods and those concerned mostly with direct and complex interactions between human agents.

Historically, CBSs have, as we have seen, taken shape around the tangible objects of manufacture, but also of retail and distribution. Corporations such as Walmart, Amazon, and FedEx, classified by economists as service companies, are in fact quasi-industrial corporations at the heart of the CBS world as businesses engaged in the distribution and sale of goods. Along with Walmart and FedEx, other pioneers of CBSs are corporations such as Toyota, Nissan, Dell Computer, and UPS. In such businesses where the speed and accuracy of production and distribution are the major determinants of success, the capacity of CBSs to measure virtually everything that happens in factories, warehouses, and depots can be the opening shot in a campaign to find ways of doings things even faster and more accurately.

But the very success of CBSs within this home base has led management theorists, system designers, and CEOs to push forward into the domain of what I will call core services, where the focus is not on the manufacture, movement, and sale of goods, but on the occurrence of complex transactions between human agents, as in financial services, health care, education at all levels, human resource management, and, as we have seen, customer relations management. Involved here is *misindustrialization*. There are two semantic variants of the concept of *industrialization* that have entered the English language: *deindustrialization,* when an economy loses major segments of its manufacturing base, as the United States did at an accelerating pace in the 1970s and early 1980s, and *reindustrialization,* the strategy attempted by the Obama administration during its second term of rebalancing the economy in favor of manufacturing.

But what is misindustrialization, especially in the context of CBSs? It has been the success of CBSs in increasing productivity in the industrial sectors of the economy that has led to their introduction in service sectors, where their use is problematic. So CBSs are expansionist technologies, and the frontier between those sectors of the economy that fall within their domain and those that do not is constantly shifting as the domain expands. There is, therefore, a need to monitor this moving frontier, just as there was a need to monitor the shifting of the western and eastern fronts during World War II.

As in the industrial sectors, the ability of CBSs to extend their methods of rigorous monitoring and control to the white-collar world depends critically on two kinds of knowledge that the systems can provide: the knowledge of whether key performance indicators are being met, and if they are not, why not. In their literature the CBS designers and the management consultants who help market their products are insistent that the power of CBSs to drill down and monitor the minutiae of production in real time, so critical to their success in an industrial context, is equally applicable to core services.

But in transferring the methods of industrial CBSs to the more complex world of core services, the system designers have had to struggle with a major obstacle. The information pyramid of industrial CBSs pivots on the system's ability to measure precisely the movements of the components, commodities, packages, and finished goods that populate the system as they move between machine shops, assembly lines, warehouses, and retail stores. It is the visibility of processes built around these objects that enables the systems to drill down and discover quickly and in real time why performance lags.

How can this regime of precise measurement and of panoptic managerial vision be transferred to a context where the objects of production are the treatment of sick patients, the transactions between teachers and pupils, or the decisions to hire and fire employees? The answer is that the structure and context of these activities must be expressed in a form that can be captured by the system, so that their digital representations can then be read and analyzed. But the limits of "capturability" become apparent when one looks at transactions between human agents where attempts to impose "capturability," and with it the disciplines of CBSs, distort the meaning of what is being done and leave the data thus generated highly vulnerable to GIGO—garbage in, garbage out.

A striking example of this expansionism is the project of reengineering the hiring side of human resource management, described in *Organizing Business Knowledge: The MIT Process Handbook* (2003).[1] This text, perhaps more than any other, gives a strong sense of Business Process Reengineering as no mere management fad but an enduring commitment of the corporate world. The handbook is the physical embodiment of MIT's online database of five thousand business processes, covering everything from human resource management to book publishing and running a restaurant. The moving force behind the handbook and the database has been Thomas W. Malone, a professor of management at the Sloan School at MIT, formerly director of MIT's Center for Coordination Science, and now director of the rather Orwellian-sounding Center for Collective Intelligence.

The project is at the intellectual pinnacle of the CBS world, bringing together leading computer scientists, system designers, management theorists, and experts in artificial intelligence and so pointing the way forward for the rank and file of the corporate world. The project's sponsors have included EDS, Boeing, Intel, UBS, the Defense Advanced Research Projects Agency, and the Defense Logistics Agency. The participation of these Pentagon agencies is significant because it testifies to the long-standing interest of the US military in military versions of CBSs as agents of the "automated battlefield." Former secretary of defense Donald Rumsfeld's enthusiasm for the military versions of CBSs significantly influenced the configuration and training of the US Army for the Second Iraq War. Despite this multidisciplinary provenance, from the outset a relentless industrial vocabulary pervades the MIT discussion of human resource management.

The familiar objective of the MIT researchers is to break down the process of "hiring an employee" into a sequence of subprocesses that can then be expressed in a form that the CBS can read, thus providing executives with a real-time, panoptic view of the corporation's HRM performance, along with the power to drill down and find the sources of error when key performance indicators are not performing as they should. But this is a more formidable task than the one faced by Walmart managers when they have to figure out how best to capture the movement of consignments of dog food between factory, warehouse, and store.

The chief problem faced by the MIT designers was that, once they started thinking seriously about the "hiring process," the number of relevant subprocesses began to proliferate, along with the number of possible ways in which the subprocesses could be combined. The designers identified six generic processes common to most hiring projects: "identify staffing needs," "identify potential sources," "select human resources," "make offer," "install employee," and finally "pay employee." But they also identified a further forty-one subprocesses, eleven for "install employee" alone: for example, "install in job environment," "install in learning environment," "install by special trainer," "install by oneself," "install before work," and "install during work."[2]

One way to deal with this proliferation of processes and subprocesses would be to allow the HR employees to rely on their judgment and experience, shaping the subprocesses to fit the particular circumstances of the person to be hired or even abandoning the whole "process" format altogether. But this concession to employee expertise blocks the ability of the CBS to capture and monitor the details of work performance. From a reengineering perspective, these self-directed employee activities are an opaque wilderness lying beyond the zone of CBS capturability and can be made sense of only in the context of an unstructured, uncapturable, and time-wasting debriefing between supervisors and subordinates.

Engineers instead have come up with what they call a "process recombinator" that allows the corporation to order and reorder the processes and subprocesses of human resource management. Their agents in this reengineering are the process designers who use the CBS software to embed corporate preferences in the detailed operations of the system. The workings of the recombinator are exceedingly complex, an inevitable consequence of the engineer's attempts to juggle the hundreds of possible combinations of processes and subprocesses and to anticipate the contingencies that might arise in an activity so subject to the vagaries of human nature and performance as human resource management.

The Rube Goldberg quality of the recombinator is captured by allowing the system designers to speak for themselves in their singular language.[3] The recombinator has three offsprings. There is a "subactivity recombinator" that "generates all possible combinations of specializations of the subactivities in the process." There is a "dependency recombinator" that generates "different combinations of coordination mechanisms for the process dependencies." Finally, there is a "bundle recombinator" that generates "different combinations of the alternatives in the dimensions represented as a bundle." This hyperelaborate apparatus of control once again bears the strong imprint of Frederick W. Taylor and Scientific Management, as it is the expertise of senior executives and system designers whose precepts about how "hiring an employee" should be done become embedded in the system's enforceable rules.

Another potent weapon of top-down control in HRM is the automated personality tests used by corporations to evaluate the suitability of their prospective employees. Of the six "generic HR processes" identified by the MIT designers, the automated personality test clearly belongs to the third, "select human resources." In 2004 eighty-nine of the Fortune 100 companies used one such test, the Myers Briggs test. Another automated test, the Wagner Enneagram Personality Style Scale (WEPPS), was used by, among others, AT&T, Boeing, DuPont, General Motors, Hewlett-Packard (HP), Proctor and Gamble, Motorola, Prudential Insurance, and Sony. Both tests take the form of a computerized questionnaire containing multiple-choice questions that the human resource specialist puts to the prospective employee, with the specialist ticking the appropriate multiple-choice boxes as the candidate responds.

Author and journalist Barbara Ehrenreich took the WEPPS test as part of her research for her book *Bait and Switch: The (Futile) Pursuit of the American Dream.* A diagram of how the test worked showed a series of interlocking triangles and circles that made Ehrenreich dizzy just to look at. The test comprised two hundred multiple-choice questions asking her such things as whether she was "sometimes," "never," or "always" special, judgmental, procrastinatory, principled, or laid back. The test revealed that Ehrenreich was Original, Effective, Good, and Loving, but that she was also Melancholy, Envious, and Overly Sensitive. The test concluded that she probably did not write very well and should attend "intensive journalistic workshops to polish her writing skills."[4] GIGO with a vengeance!

THE INDUSTRIALIZATION OF human resource management brings the disciplines of standardization, measurement, and speed to an activity already strongly rooted within the corporate world. The most egregious example of misindustrialization I have yet come across goes far beyond this world and brings us to the high medieval and Renaissance palaces of the University of Oxford, of all places. Here the agent of misindustrialization is the academic production regime that has enveloped the university, along with all the British universities, during the past twenty-five years, a regime heavily influenced by American management systems, with their regimes of rigorous quantification, their proliferation of key performance indicators, and their omnipresent apparatus of monitoring and control.

Outside of the UK's own business schools, not more than a handful of British academics know where the management systems that so dominate their lives come from and how they have ended up in Oxford, Cambridge, Durham, and points beyond. The most influential of the systems began life, as we have seen, with IT corporations such as IBM, Oracle, and SAP; moved eastward across the Atlantic by way of consulting firms such as McKinsey and Accenture; and reached UK academic institutions through the agency of the UK government and its satellite bureaucracies. Of all the management practices embraced during the past twenty-five years by IT system providers (IBM, Oracle, SAP), consulting firms, and business schools, the one that has had the greatest impact in British academic life is also among the most obscure, the Balanced Scorecard.

The BSC is the brainchild of Robert Kaplan, an academic accountant at the Harvard Business School, and Boston consultant David Norton. On the seventy-fifth anniversary of the *Harvard Business Review* in 1997, its editors judged the BSC to be among the most influential management concepts of the *Review*'s lifetime.

Kaplan and Norton have promoted their concept in eight *Harvard Business Review* articles, beginning with "The Balanced Scorecard: Measures That Drive Performance" (January 1992), "Putting the Balanced Scorecard to Work" (September 1993), and "Using the Balanced Scorecard as a Strategic Management System" (January 1996). As befits Kaplan's roots in accountancy, the methodologies of the BSC focus heavily on the setting up, targeting, and measurement of statistical key performance indicators. In their 1992 piece, Kaplan and Norton classify the KPIs of the Balanced Scorecard under four headings: financial performance, internal business processes, innovation and learning, and customer service.

This multiplication of key performance indicators also multiplies the opportunities for top-down monitoring and control, and Kaplan and Norton indeed use the language of aviation and the autopilot to describe the BSC at work: "Think of the balanced scorecard as the dials and indicators in an airplane cockpit," with the CEO, his senior executives, and their system designers in control. That the tentacles of the BSC should have crossed the Atlantic and enveloped the great and ancient university of Oxford, founded in the early thirteenth century and a wellspring of the Western humanist tradition ever since, that Oxford of all places should be subject to misindustrialization, testifies to the metastasizing powers of the CBS world and its capacity to threaten humanist values embedded in institutions located far beyond the frontiers of the market economy, and also far beyond the frontiers of the United States.

For me this misindustrialization of Oxford has been personal history. In 2005 I returned to the university as a late-career academic, forty years after my time there as an undergraduate. My dawning awareness of what had been happening during these intervening decades was a slow-motion version of the experience of seeing Harold Pinter's late play *Party Time*, the story of how the rituals of bourgeois social life continue on their way as a totalitarian darkness descends around them.[5] Pinter's late plays became shorter, increasingly political, and preoccupied with the psychological menace not so much of solitary misfits, as in early plays such as *The Homecoming* and *The Caretaker*, but of whole social classes, and especially of the London bourgeoisie empowered by Margaret Thatcher from the 1980s onward.

In *Party Time*'s opening scene, a group of what seem to be London businessmen and financiers is sitting around in their luxury apartment, celebrating their latest material acquisitions, putting down their female companions, and treating all those who come into contact with them with contempt. However, amid this (for Pinter) routine nastiness, his characters let slip comments and allusions that point to the presence of something much more sinister. There is talk of "a bit of a round up this evening," which is coming to an end so that "normal service will be resumed shortly."[6] The comments are made very much *en passant* because Pinter's heroes don't want to alarm their female companions and spoil the evening's fun, but it gradually dawns on us, the audience, that these are not businessmen at all but operatives of a British Stasi; that Britain has become an authoritarian, fascist state; and that violence and torture are in the air when there's trouble downtown and the regime deals with its opponents.

In invoking this precedent for my encounter with academic misindustrialization at Oxford, I am not suggesting that Britain is succumbing to fascism or that the agents of academic managerialism at the university are heirs of the Stasi. It was the manner of my becoming aware of recent Oxford history that brought to mind Pinter's play. Among many of my academic contacts, there was a reluctance to provide a full-blooded account of this history. As in *Party Time,* the system made its presence felt through fleeting allusions that only gradually sorted themselves out into something that made sense, and this reticence suggested a certain deference in the face of bureaucratic power.

When describing their day-to-day scholarly lives, my academic contacts used a strange and, in an academic context, unfamiliar language. They spoke of "departmental line managers" who monitored their work. They speculated whether an academic conference they were going to attend would count as an "indicator of esteem." They referred to academics of no great distinction who had been given personal professorships as a reward for their steady output of books, albeit of uneven quality, and to academics from overseas, especially the United States, who had become temporary members of a department, valued for their output of books. Then in 2006 and 2007 the acronym *RAE,* standing for *Research Assessment Exercise,* began more and more to feature in these remarks. This was the academic production regime, mandated by all UK governments from Margaret Thatcher's time onward and requiring academics to turn out a designated number of books, monographs, or articles in learned journals over a four- or five-year period.

So Dr. X, an academic with a sparse publication record, had not been entered for the RAE by his department, and Dr. Y was in danger of missing his book deadline for the RAE with possibly dire consequences, and the Department of Z, which had not gotten the top grade in the last RAE, was in danger of doing so again and was being hounded by the university administration as a result. What was going on here? It slowly became apparent that much academic life at Oxford was taking place in the shadow of an elaborate system of bureaucratic command and control, put in place by Margaret

Thatcher's Conservative government in the late 1980s, maintained and enlarged by all its success, whether Conservative or Labour, and profoundly influenced by management practices such as the Balanced Scorecard that had originated in the United States.

The intervention of the British state in the management of academic research has created layers of bureaucracy, linking the UK government at the top all the way down to the scholars at the base—researchers working away in libraries, archives, and laboratories. In between are the bureaucracies of HEFCE (the Higher Education Funding Council for England), of the central university administrations, and of the departments of the universities themselves. HEFCE itself is a special state bureaucracy, situated between the government and the universities and set up by the government to handle the detailed administration of the system.

HEFCE's control regime, via the RAE, is an example of quantification and control applied to higher education, serving one of government's chief objectives: to make the universities more like business in the way they conduct their affairs and to give business a greater role in the shaping of academic research. In the words of David Lammy, minister for higher education in the Labour Government of Gordon Brown (2007–2010), but in words that could have been uttered by any of his predecessors or successors of the past twenty-five years: "We propose that the panels assessing [research] impact will include a large proportion of the end-users of research—businesses, public services, policymakers and so on—rather than just academics commenting on each other's work."[7]

The HEFCE control regime relies on a proliferation of key performance indicators whose targets must be met if a university department's research is to be funded. This has led to the measurement and targeting of a scholar's research output and measurement of the time taken by the scholar to do the research, the money the research brings in, the indexes of impact and esteem surrounding the research, and the grades awarded for research submitted in the RAE, by panels of experts set up by HEFCE.[8] With 52,409 academics entered for the most recent RAE of 2008, more than 200,000 items of scholarship reached HEFCE. For the previous RAE of 2001, the avalanche of academic work was so great that it had to be stored in an unused aircraft hangar located near HEFCE's headquarters in Bristol, England.

With each RAE the incoming items of scholarship are examined by academics on panels set up by HEFCE to cover every discipline from dentistry to medieval history—sixty-seven in the 2008 RAE. Each panel is usually made up of between ten and twenty specialists selected by their respective disciplines, though subject at all times to HEFCE's rules for the RAE. The panels had to award each submitted work one of four grades, ranging from 4*, the top grade, for work "whose quality is world leading in terms of originality, significance and rigor," to the humble 1*, "recognized nationally in terms of originality, significance, and rigor."

The HEFCE control system is simply a typical corporate Balanced Scorecard dressed up in an unfamiliar language. Writing in January 2010, British biochemist John Allen of the University of London described how this blizzard of targets and metrics descended upon his professional life: "I have had to learn a new and strange vocabulary of 'performance indicators,' 'metrics,' 'indicators of esteem,' 'units of assessment,' 'impact' and 'impact factors.'"[9] One might also mention tallies of medals, honors, and awards bestowed ("indicators of esteem"); the value of research grants received; the number of graduates and postdoctoral students enrolled; and the volume and quality of "submitted units" of research output.

There is one significant difference between an authentic Balanced Scorecard and the hybrid version imposed on British universities by the UK government, via HEFCE. In the corporate model, the entire apparatus of targeting, monitoring, diagnosis, prescription, and sanction belongs to the corporate mother ship. But in the hybrid British model, HEFCE performs only the first, targeting, function, and the four other functions are devolved to university divisions, departments, and line managers, who carry out the monitoring, diagnostic, prescriptive, and sanctioning tasks on HEFCE's behalf. Here is a description of what it is like to be at the receiving end of the HEFCE/RAE system, from a young and very promising historian working in one of the newer universities in the London area:

> The bureaucratization of scholarship in the humanities is simply spirit-crushing. I may prepare an article on extremism, my research area, for publication in a learned journal, and my RAE line manager focuses immediately on the influence of the journal, the number of citations of my text, the amount of pages written, the journal's publisher. Interference by these academic managers is pervasive and creeping. Whether my article is any good, or advances scholarship in the field, are quickly becoming secondary issues. All this may add to academic "productivity," but is it worth selling our collective soul for?

In the twenty-five years the HEFCE system has been in existence, the values of the HEFCE control system have become internalized among many of its subjects in an academic variant of Bentham and Foucault's panopticon, fortified by the old British imperial strategy of divide and rule and also by the techniques of the Toyota production system, whereby the failings of any one member of the production team become the collective responsibility of the whole team, with the collective resentments of the team adding to the shame of the offending worker—a description that closely fits what can go on in a university department in the run-up to the RAE deadline.

5

THE CASE OF GOLDMAN SACHS

IN THE FINANCIAL CRISIS OF 2007–2008, CBSs PERFORMED ON A much bigger stage than any we have encountered so far. The scope and impact of the systems extended beyond the corporate and the national to the global, and the damage inflicted was correspondingly great. In the financial crisis, CBSs and their constituent technologies came together with an unprecedented malignancy. The operations of Wall Street's mortgage machine before and during the crash, and the role of CBSs within the machine, closely fits what Joseph Schumpeter called the "mechanization of progress," whereby innovation becomes "depersonalized and automated" and "bureau and committee work" replaces individual actions and judgments.[1]

The Wall Street machine relied on information technologies to create a virtual assembly line on which something as simple as a single subprime mortgage at the start of the line could become by the time it reached the end a molecule within a financial derivative so complex that it was beyond the powers of the IT systems themselves to manage or keep track of. Amid these highly complex IT systems, it was easy to forget that this vast, inverted pyramid of financial manipulation pivoted on the creditworthiness of countless middle- and lower-income families obtaining mortgages for the first time for homes they could ill afford.

As with the making of Ford's Model T, the making of a financial derivative moved through multiple stages, with each stage responsible for adding an essential component to the product. On the Wall Street line, in contrast to Ford's, these way stations were also independent financial agencies, each exacting hefty fees and markups as the product passed through its segment of the line. The US government—J. K. Galbraith's "countervailing power"—which might have set limits on the machine's operations, was in fact actively working on the machine's behalf. In their book *Thirteen Bankers,* Simon Johnson and James Kwak show in detail how the regulatory regime that allowed the machine to run amok was as much the work of Wall Street Democrats such as Robert Rubin and Larry Summers as it was of Reagan Republicans such as Donald Regan and Senator Phil Gramm of Texas.[2]

As the machine worked flat-out in the run-up to the crisis, there were eight principal businesses at work along the line: the mortgage brokers who worked directly with the subprime clients; the mortgage bankers, who, benefiting from the recommendations of the brokers, underwrote the subprime mortgages, bundled them together, and passed them on to investment bankers as mortgage-backed securities (MBSs); the mortgage servicers responsible for collecting the monthly mortgage payments from the subprime clients, even as the ownership of the mortgages moved from one remote owner to another along the line; the investment bankers who bundled the MBSs with further bundles of debt—student loans, credit card debt—to form collateralized debt obligations (CDOs); the rating agencies—Moody's, S&P, Fitch—who examined the CDOs and, with a stroke of financial alchemy worthy of Merlin the magician, transformed CDOs heavy with poorly rated MBSs into top-rated derivatives with a AAA rating; the insurers, such as AIG, who made it possible for anyone, whether they owned a CDO or not, to take out an insurance policy—a credit default swap (CDS)—against a CDO's possible default; the CDO brokers who marketed the newly sanitized derivatives; and at the very end of the line the purchasers of the CDOs and MBSs beyond Wall Street—foundations, universities, pension funds, midwestern school districts, German regional banks, all very big losers once the MBSs and CDOs went bad.

The millions of mortgages, student loans, and other debt contracts that constituted the raw material of the machine had a dual existence both as documents in the safekeeping of legal custodians, themselves a minor component of the machine, and as electronic bits in digital space. Once the transformation from the physical to the digital had taken place, the transactions could be assembled, subassembled according to risk, and moved between way stations at high speeds.

As we have seen, speed is also achieved in mass production by automating as far as possible the cognitive functions that may have to be performed at points along the line. That is why call-center agents deal with their clients with the use of digital scripts and why HMOs are constantly pressing their physician clients to observe standard treatment protocols that, according to the HMO database, are faster and cheaper than the alternatives. Unless this happens, and if instead employees are allowed to exercise their own judgment free of rigorous time constraints, then the business process or subprocess will not achieve management's targets for time and cost—its key performance indicators—and, from management's perspectives, the system will have failed.

The Wall Street machine was able to achieve the speed that it did only by automating, at three critical points along the line, complex judgments about financial instruments that should have been subject to painstaking, time-consuming analysis. But once again, the rules of this automated decision making were always the outcome of executive decision making that, once embedded in the system, had to be followed by front-line employees. The Wall Street machine was therefore as much an example of digital industrialism as the call centers of the front office. The operatives of the Wall Street machine relied on software-born indexes of risk to pass favorable judgment on the derivatives as they moved along the line.

The first of the three indexes at the heart of the crisis was the FICO score, used to estimate the creditworthiness of mortgage borrowers, that could be gamed by system designers to show that a Mexican immigrant worker with an income of twenty thousand dollars could handle a subprime mortgage worth three hundred thousand dollars. The second of the indexes was the rating agencies that treated subprime borrowers as if they were small businesses and, looking at the historical record for small business failures, found that the probability of subprime default was low. Finally, the value-at-risk (VAR) indexes pioneered by Professor Philippe Jorion of the University of California were heavily relied on to assess the risk of CDOs. For its failure to allow for the unexpected and exceptional, Nassim Taleb of the *Black Swan* characterized this index, a decade before the debacle, as "charlatanism" and "potentially dangerous malpractice" for a "school for sitting ducks."[3]

WE WILL NOW, in the language of CBSs, *drill down* and look at how integration within the mortgage machine shaped the conduct of one of Wall Street's leading actors, Goldman Sachs. Goldman's handling of the Wall Street crash during its critical, formative months between the middle of 2006 and the end of 2007 must be among the most heavily documented events in modern business history. Pride of place in this bibliography goes to the Senate Permanent Subcommittee on Investigations' (the Levin Committee's) 266-page report on Goldman, "Failing to Manage Conflicts of Interest: A Case Study of Goldman Sachs," which is just one section within a 639-page report on the role of investment banks in the crisis.[4] The report draws on the tens of thousands of e-mails subpoenaed from Wall Street firms by the committee and provides a day-to-day, and sometimes hour-by-hour, account of what went on at Goldman during those months. A companion volume to the report is the transcript and video footage of the hearings before the Levin Committee, which took place on October 27, 2010, when Goldman's crisis team, from CEO Lloyd Blankfein down to the humblest traders, gave their side of the story.[5]

In addition, there is the growing list of lawsuits against Goldman, all with their accompanying texts, brought by government agencies such as the Securities and Exchange Commission (SEC) and the Federal Housing Authority, and also by aggrieved clients of Goldman such as the now defunct Australian hedge fund Basis Capital and by ACAS Capital, which collaborated with Goldman in the creation of the Abacus CDO, notorious for the role of the hedge-fund manager John Paulson in selecting the securities to be included in the CDO. In July 2010, the SEC fined Goldman $550 million to settle charges that it "failed to disclose to investors vital information about the Abacus CDO . . . particularly the role that Paulson and Co. played in the portfolio selection process." In other court actions against Goldman the actions are still pending and Goldman's level of culpability has yet to be decided. But the Levin Committee's report and hearings provide, I believe, strong evidence that the deception and manipulation of clients eventually became an integral part of Goldman's trading strategy. [6]

In trawling through the documents, it is essential never to lose sight of the role of the Goldman corporate hierarchy in the crisis and the highly disciplined way in which it managed the company. The answer to the old Watergate question "What did the president know, and when did he know it?" is, in the case of Goldman's big three—Lloyd Blankfein, CEO; Gary Cohn and Jon Winkelried, copresidents—that they knew everything that mattered (and in real time). The big three's point man for the crisis, Goldman's own H. R. Haldeman, was copresident Gary Cohn, whose name turns up frequently in the e-mail flow of his subordinates. Although the e-mail vocabulary between Blankfein, Cohn, and their subordinates has a certain locker-room familiarity about it, there is never any doubt about who is in charge.

Blankfein, Cohn, and their team were men of the machine and, as it turned out, among the most skillful manipulators of the machine on Wall Street. By the early 2000s, Goldman's derivatives trading could no longer be called banking in any meaningful sense of the term, but had become an industrial activity, turning out virtual products whose fortunes depended on the efficient management and coordination of processes: the accumulation of mortgages and other forms of debt from bankers and brokers, their transformation into financial derivatives, and their selling on to clients.

In Goldman's culture these processes were of supreme value because they were vehicles for the creation of Goldman's earnings and profits, and so critical for the health of the Goldman stock price, the size of the salaries and bonuses paid out to Goldman executives, and the reach of Goldman's power and fame on Wall Street and beyond. As long as house prices continued to rise and those along the process chain continued to make money, the model's flaws could be ignored, notably the dismal quality of the subprime mortgages upon which the whole system depended.

But once the housing market turned, the system collapsed. The difference between Goldman and the other leading players on Wall Street was that Goldman saw it coming and was able to recalibrate its machine so that not only did it avoid the catastrophic losses that destroyed Lehman Brothers and crippled Citicorp, but it actually came out ahead. But to achieve this Goldman behaved, I will argue, with ruthless cynicism, above all in deceiving and exploiting its clients. Why did Goldman do this? The simple answer is that for Goldman, wealth creation on its own behalf took priority over everything else, and nothing was going to stand in its way. In histories that tell of how Goldman acted on its view of the deteriorating markets and came out ahead, the word *warehouse* often appears, a choice word that is revealing both as a pointer to the heavily industrial character of Goldman's trading activities and as providing a mental diagram to locate the various financial instruments Goldman was dealing in. But *warehouse* does not fully capture the reality of what was going on at Goldman.

A real warehouse is a place where finished goods are stored before they are shipped off to customers or retailers, whereas Goldman's "warehouse" was much more like a factory where industrial processing of financial instruments took place on a virtual assembly line. The Goldman "factory" was an electronic space where the "raw materials" of loans coming in from such mortgage brokers as New Century, Long Beach, and Countrywide were processed on the virtual line into financial instruments such as mortgage-backed securities, collateralized debt obligations, and credit default swaps, which then were marketed to clients. In the precrisis world, where it was assumed that house prices would go on rising indefinitely, the processes of "securitization," that is, the processing of the raw loans coming in from the brokers, were relatively straightforward.

In the case of mortgage-backed securities, the incoming loans were bundled together by Goldman, their ownership vested in a trust, with the trust issuing securities to investors, which gave them the right to the cash flows generated by the loans as householders paid off their mortgages. With collateralized debt obligations, several MBSs would be bundled together, along with bundles of student, consumer, or corporate loans, again with ownership of the loans vested in a trust, with the trust issuing securities to investors. In addition, there was a whole superstructure of insurance, known as credit default swaps, attached to the MBSs and CDOs. The owners of the MBS and CDO securities, or indeed anybody, could take out an insurance policy against their loss of value and receive compensation if this happened. Equally, the owners of the securities, if they were confident that they would hold their value, could be the providers of insurance and receive regular premium payments from the policy holders. If the securities lost value, then, as with any insurance policy, the issuer of the insurance was obliged to compensate the policy holder for their loss.

Whether Goldman was pursuing a coherent trading strategy during these crisis months is a fraught issue, because it is claimed by those suing Goldman that it had a consistently pessimistic view of the market, and although this pessimism shaped its trading on its own behalf, Goldman hid this pessimism from many of its clients, fed them an upbeat view of the market that it did not believe in, and persuaded them to buy assets that it knew were flawed and would lose value, and indeed did. Goldman's defense all along has been that it had no aggressive moneymaking strategy at all and was simply the prudent guardian of its clients and its own interests. In its own words, "The risk management of the firm's exposures and the activities of our clients dictated the firm's action, not any view of what might or might not happen to any security or market."[7]

This was also the line taken by Lloyd Blankfein himself during his appearance before the Levin Committee on October 27, 2010, where he spoke of Goldman as virtually a charitable organization, the passive counterparty in deals where strong-minded clients came in and told Goldman exactly what they wanted, and Goldman respectfully executed their instructions. So Blankfein: "The customers who are coming to us for risk in the housing market wanted to have a security that gave them exposure to the housing market, and that is what they get. . . . [T]he security itself delivered the specific exposure that the client wanted to have." And again: "What clients are buying, or customers are buying, is they are buying an exposure. The thing that we are selling to them is supposed to give them the risk they want. They are not coming to us to represent what our views are. . . . [T]he institutional clients we have wouldn't care what our views are. They shouldn't care."[8]

At the Senate hearings the Goldman team from Blankfein on downward was much helped in its own defense by the extreme complexity and variety of the derivatives Goldman traded during the crisis period and the difficulty for laymen, including those on the Levin Committee, of distinguishing between them, and especially the differing legal obligations attached to each of them. It was here that the Levin Committee hearings fell short in ways that undermined the impact of the report and the hearings in the public policy debate. In his preamble to the hearings, Senator Carl Levin (D-MI) drew attention to Goldman's differing obligations to its clients as market maker and underwriter, but Levin and his fellow senators lost sight of this distinction when questioning Blankfein and his colleagues, allowing them to slip away again and again behind a fog of obfuscation.

One way of cutting through this obfuscation is to imagine for a moment that Goldman was a real industrial company making and selling real products, rather than a virtual industrial company making and selling virtual products. Imagine Goldman for a moment as the Goldman Motor Manufacturing Company, or GMMC, a Detroit competitor of Ford in the early days of mass production in the 1920s. One day GMMC discovers to its horror that there is a serious flaw in its manufacturing processes so that a significant percentage of the engines installed in its best-selling model, the Model G, break down after just a few weeks on the road. The vice president for manufacturing tells the CEO that GMMC must immediately close its plant for retooling, recall the products it has already sold, and strip down the models it has in stock, selling off the uncontaminated engine parts to local component dealers.

But the vice president for finance quickly does his sums and persuades his colleagues that the cost of this plan A is too great, and they must decide instead on plan B. With plan B, GMMC instructs its salesmen to avoid its local Detroit dealers, whom the company suspects have gotten wind of problems at the plant, and tells the GMMC salesmen instead to ship the problem Model Gs to the South and sell them directly to unsuspecting farmers in rural Kentucky and Tennessee. Without telling these customers that there is anything wrong with the Model Gs, GMMC quietly takes out an insurance policy with a local Detroit company that will pay out to GMMC every time one of its cars goes wrong. When the owners of the stricken vehicles demand a refund, GMMC refuses. This is a simplified but essentially accurate account of what Goldman frequently did in its derivatives trading. Looked at day-to-day, Goldman's trading strategies were complex, sometimes counterintuitive, and lacking in obvious direction. The GMMC fable can be a helpful guide as we try to make sense of what Goldman was doing.

DRAWING ON ITS vast e-mail trove, the Levin Committee report shows Blankfein's image of Goldman as a charitable organization to be entirely fictitious and repeatedly quotes chapter and verse to prove it. The report shows that from late 2006 onward, Goldman's senior executives had a consistently pessimistic view of the housing market and the financial instruments attached to it and thenceforth pursued an aggressive trading strategy to maximize its gains from the crisis, with the manipulation of clients becoming, I will argue, an integral part of Goldman's chosen strategy. The evidence of how Goldman's top executives really viewed the markets is therefore germane to this whole history, and some of it now follows.

In an internal "self-review" dated September 26, 2007, Michael J. Swenson, head of the Goldman Sachs Mortgage Department's Structured Products Group, wrote that "during the early summer of 2006 it was clear that the market fundamentals of subprime and the highly leveraged nature of Collateralized Debt Obligations were going to have a very unhappy ending." On December 7, 2006, Daniel Sparks, head of Goldman's Mortgage Department and a key link between the trading floor and the Goldman big three, exchanged e-mails with Thomas Montag, a senior Goldman executive and co-head of Global Securities for the Americas, "about why Goldman was not doing more to reduce the firm's risk associated with its net long positions" in housing-related assets.[9]

On December 14, 2006, David Viniar, chief financial officer and therefore number four in the Goldman hierarchy after the big three, convened a meeting in the conference room next to his office on the thirtieth floor (the seat of power where the big three also had their offices), where they conducted an in-depth review of the Mortgage Department's holdings because its "position in subprime mortgage related assets was too long, and its risk exposure was too great."

The next day Viniar e-mailed Montag about the deteriorating markets and the opportunities it opened up: "My basic message was lets [*sic*] be aggressive distributing things because there will be very good opportunities as the markets [go] into what is likely to be even greater distress and we want to be in a position to take advantage of them."[10]

Then on February 11, 2007, and from the thirtieth floor itself, Blankfein urged the Mortgage Department to get on with the task of selling off its deteriorating assets: "Could/should we have cleaned up these books before and are we doing enough right now to sell off cats and dogs in other books throughout the divisions?" On February 14, 2007, Daniel Sparks reported further on the deteriorating markets and the trading opportunities it opened up: "Subprime environment—bad; and getting worse. Everyday [*sic*] is a major fight for some aspect of the business. Credit issues are worsening on deals and pain is broad. . . . [D]istressed opportunities will be real, but we aren't close to that time yet."[11]

In 2006 and 2007 Goldman originated twenty-seven CDOs and ninety-three MBSs with a total value of about $100 billion.[12] The problem for Goldman from the summer of 2006 onward was that its "factory" was clogged with components of MBSs and CDOs in varying stages of manufacture—raw loans just in from the brokers and not yet bundled, loans in the process of being bundled, bundles of MBSs not yet put together as CDOs, finished CDOs and MBSs not yet marketed, and securities of old CDOs and MBSs that Goldman had not yet sold and were still in the factory. Goldman had acquired and processed these assets on the assumption that the housing market was strong and that there would be an equally strong demand for the finished CDOs and MBSs.[13] But now the market was about to collapse, and Goldman had billions' worth of what it believed would become failing assets on its hands. So what to do with them?

Much of what Goldman then did was what any owner of a big stock portfolio would do if faced with a collapsing market. Goldman stopped taking in any more loans from the mortgage dealers; it abandoned some CDOs and MBSs that were still "under construction" and liquidated others that were fully formed, selling off their components in the markets, as it also did with some of the "raw" loans recently acquired from the brokers that had yet to enter the securitization process. If this is all that Goldman had done, there would be no Goldman story, Blankfein would be esteemed on Wall Street as the great survivor, and Goldman would not be the target of multiple lawsuits that could still cost it hundreds of millions, if not billions, of dollars.

But what Goldman also did, and this has been the source of its troubles, was to persist with the creation of new CDOs and MBSs and to continue with the marketing of existing ones, even bringing some of its own proprietary assets into the factory so that they too could become part of a new CDO. Goldman also began taking the insurance structure pivoting on the CDOs, the credit default swaps, much more seriously as a potential source of revenue. Although the trading strategies involved in these activities were sometimes complex, the motive underlying them was simple and straightforward. Goldman believed that it could make more money by disposing of its factory assets as components of CDOs, MBSs, and CDSs than by just selling them off unadorned in the market.

The problem that then arose for Goldman was that in marketing the three kinds of financial derivatives, the company was acting as underwriter and placement agent and not simply as market maker or trader on its own behalf. As underwriter and placement agent, Goldman was subject to rules on fair disclosure that as market maker it was not. In the Levin Committee hearings, the Goldman team, from Blankfein on down, went to very considerable lengths to blur the distinction between the different roles and to cast themselves, whenever possible, as humble market makers. There can be little doubt that in preparing for the hearings, a high-risk event for Blankfein et al., Goldman's extremely high-priced lawyers got together with their clients and advised them that, in view of their record, obfuscation on the distinction between underwriter and market maker was advisable.

This evasion was much in evidence at the hearings and especially in the gladiatorial contest between Senator Levin and Blankfein on the final day, when Levin searched with increasing frustration for the smoking gun that would sink Blankfein but never quite managed to find it. In reading the committee transcript, and simultaneously watching the contest on a podcast available at the Levin Committee website, I was reminded not so much of Watergate and the search for the Nixonian smoking gun as a remarkable scene that appears in several versions on YouTube.

In it a gigantic Alaskan brown bear sits on a slab of rock overlooking a fast-moving river full of salmon. The bear is trying to scoop up one of the salmon with his paws, but most of the time the salmon are much too quick for him and he fails. But just occasionally, he succeeds and has a terrific meal. There was something bearlike about Senator Levin peering down at Blankfein from his senatorial perch as he flailed around, trying to land his quarry, but with the slippery Blankfein swimming clear every time. The blurring of the distinction between market maker and underwriter was central to Blankfein's evasive strategy, as the following exchanges reveal:

SENATOR LEVIN: Is there not a conflict when you sell something to somebody and then are determined to bet against that same security and you didn't disclose that to the person you are selling it to? Do you see a problem?

LLOYD BLANKFEIN: In the context of market making, that is not a conflict. What clients are buying, or customers are buying, is they are buying an exposure. The thing that we are selling to them is supposed to give them the risk they want.[14]

Senator Levin wasn't satisfied:

SENATOR LEVIN: How about you are investing in these securities. This isn't a market making deal. This is where you have a decision to bet against, to take the short side of a security that you are selling, and you don't think there is any moral obligation here?

LLOYD BLANKFEIN: Every transaction Senator, and this is—and I think it is important and again, I am not trying to be resistant but to make sure your terminology—when as a market maker, we are buying from sellers and selling to buyers . . . [15]

Levin cuts him off but later returns to the attack:

SENATOR LEVIN (with increasing frustration): You are betting against that same security you are out selling. I have just got to keep repeating this. I am not talking about generally in the market. I am saying you have got a short bet against that security. You don't think the client would care?

LLOYD BLANKFEIN: I don't Senator. I can't speak to what people would care. I would say that the obligations of a market maker are to make sure your clients are suitable and to make sure they understand it. But we are a part of a market process. We do hundreds of thousands, if not millions of transactions as a market maker.[16]

Again the Uriah Heep side of Blankfein as he relegates Goldman to being just another humble market maker along with all the others on the trading floor at the New York Stock Exchange. This was a clever strategy, but also high risk. It was clever because anyone with a serious interest in the stock market, which presumably included most of the members of the Levin Committee, knew what market makers do. They hold a supply of a stock for those who wish to buy or sell it. They adjust the price with shifts in supply and demand, and they are obliged to buy and sell at prices broadly in line with the rest of the market. It is not their role to advise investors about the wisdom of buying a stock, and they are not at fault if a stock loses 20 percent of its value within an hour of its purchase.

With his answers to Senator Levin, Blankfein was trying hard to cloak Goldman and himself in the passive neutrality of the market maker. But this was also high risk because it would have taken only one senator with forensic lawyerly skills to rip through this defense and reach—at last—the smoking gun. With the marketing of CDOs and MBSs to its clients, Goldman was not a market maker but an underwriter or placement agent and was therefore subject to rules of disclosure about the suitability of its product for the investor that it had manifestly violated. To grasp the sheer chutzpah of Goldman's marketing, one needs to look at one of these deals in detail. Among the most revealing was Goldman's marketing of the Timberwolf CDO between September 2006 and June 2007, the very months when its own view of the housing market soured and when it began to dispose of its own flawed assets.[17]

TIMBERWOLF WAS, in the arcane language of financial derivatives a "synthetic CDO," which meant that it comprised cash CDOs, securities that gave investors the right to receive the income flows from mortgage interest payments and from interest payments from other forms of debt held by the CDO trust such as consumer and student loans. As a "synthetic CDO," Timberwolf also comprised credit default swaps, insurance contracts in which one party offers insurance against a security's loss of value, and the other party buys the insurance and can do so even if he does not own the security. In May and June 2007, Goldman marketed Timberwolf to Basis Capital, an Australian hedge fund, and sold it $100 million worth of insurance contracts whereby Basis took the "long" side of the deal and offered insurance to counterparties who took the "short" side.[18]

As we have seen, for Basis to make money on the deal, the securities referenced by the CDS had to hold or increase their value and so yield for Basis a steady flow of premiums paid by the counterparties on the short side and with the insurance contracts Basis owed maintaining or increasing their value. If the securities did not hold their value, Basis as insurer stood to lose money, both as the resale value of its insurance contracts declined and as it paid out to those who had insured against the loss.

In making the sale to Basis Capital, Goldman relied on a pitch book that talked up the advantages of the deal.[19] The pitch book assured Basis that Timberwolf was structured to "generate positive performance for the benefit of both debt and equity investors." Also, "the objective of selecting assets and identifying reference securities was to identify and exclude transactions that contain potentially adverse features; including higher risk, lower quality." Timberwolf was structured with "an objective of zero loss for CDO debt investments."

Goldman's cash-flow analysis estimated that "Timberwolf would return positive performance and represent a secure investment." George Maltezos, the Goldman sales representative handling the Basis sale in Australia, assured the company that "we would certainly appreciate your support, and equally help create something where the return on invested capital for Basis is over 60 per cent."

In making a sale based on claims that, at the time, it had many reasons to believe were false—as the e-mail record published by the Levin Committee clearly demonstrates—Goldman really was descending to the ethical level of our fictional GMMC automaker. It was selling the equivalent of a clunker off the lot that it knew would break down within days of the sale, but hid this information from the customer. Again, the Levin Committee's e-mail hoard provides chapter and verse and shows that what happened was not the aberrant, freelance behavior of a sales agent operating very far from home in Sydney, Australia, but was a carefully thought-out strategy engaging the Goldman hierarchy in New York all the way up to the thirtieth floor itself.

Already in the early months of 2007, when Timberwolf's component assets were still being accumulated in Goldman's CDO warehouse/factory, Goldman was concerned about the falling value of these assets as the market for subprime and other housing-based assets continued to deteriorate. We have already seen that Goldman's concern about this deterioration dates from at least the summer of 2006 and that in February 2007 Daniel Sparks, head of the Mortgage Department, had characterized the "subprime environment—bad; and getting worse. . . . Credit issues are worsening on deals and pain is broad."[20]

Also in February 2007 Sparks had told Thomas Montag that "due to falling subprime prices, the assets accumulated in the warehouse account for the $1 billion Timberwolf CDO had already incurred significant losses." With such losses hitting many of its warehoused CDO assets, Goldman began liquidating those CDOs that it felt were no longer in a fit state to be marketed to clients, selling off their assets directly into the market. In an e-mail to senior executives dated February 23, Sparks estimated that Goldman had lost $72 million on holdings in its CDO warehouse accounts and told them that he had liquidated three CDO accounts worth $530 million.[21] Timberwolf, however, was not among the casualties, despite the "significant losses" mentioned by Sparks in his February 23 e-mail to Montag.

Thus, once Timberwolf "closed" on March 27, 2007, three weeks ahead of schedule, and now with its full inventory of securities nominally worth $1 billion, Goldman faced the problem of how to market a product that it already had strong grounds for thinking was flawed and also knew that the longer sales were delayed and the CDO securities remained sitting in the Goldman warehouse/factory, the greater the scope for further falls in values and further losses for Goldman. Goldman therefore gave high priority to Timberwolf sales. In an e-mail sent to senior Goldman executives three weeks in advance of the Timberwolf "closure," Sparks told them, "I can't over state the importance to the business of selling these positions and new issues."[22]

As early as February the Goldman sales force had been working on a list of clients to target for Timberwolf sales, and during the spring and summer the "Goldman Syndicate," a subcommittee at New York headquarters responsible for coordinating sales efforts, sent out what were known in the business as "Senior CDO Axes," high-priority sales directives distributed on a weekly and sometimes daily basis, many placing a high priority on selling Timberwolf securities and spurring on the Goldman sales force with promises of "ginormous" credits if they succeeded. "Lets [*sic*] double the current offering of credits for Timberwolf," Sparks suggested on April 19, only to be told, "We have done that with Timberwolf already."[23]

But by then sales had stalled, and by the second week of May Goldman had made only one Timberwolf sale in the previous several weeks. Meanwhile, the value of Timberwolf securities continued to fall. Goldman's response was to start targeting what it called "nontraditional" clients for Timberwolf. These clients, like Basis, were located far from New York, and their knowledge of what was happening in the subprime markets would not compare to the knowledge of Goldman's clients near at hand, who had conspicuously failed to buy Timberwolf securities in April and May. These outlying clients were the financial counterparts of GMMC's rural customers in Kentucky and Tennessee. The plan for Timberwolf was to target "institutional buyers that can take larger bite size than traditional CDO buyers . . . for example Asian banks and insurance companies."[24] Along with Basis in Sydney, other outlying clients were Bank Hapoalim in Tel Aviv, Israel, and a Korean insurance company call Hungkuk Life in Seoul.

With these sales the Goldman sales force was also subject to a full-court press from New York, which could reach all the way to the thirtieth floor. On June 13, 2007, the day that he clinched the deal with Basis, Goldman's salesman in Sydney received an anxious e-mail from Sparks, saying, "Let me know if you need help tonight. . . . I'd love to tell the senior guys on 30 [i.e., the thirtieth floor] at Risk Committee Wednesday morning that you moved $100 million."[25] Spark's urgency about the sale to Basis and his desire to bring good news to the big three on the thirtieth floor stemmed from his knowledge that the Timberwolf assets were still losing value and there was no time to be lost in getting rid of them. This information was not, however, communicated to Basis Capital, and a widening gap opened up between the price at which Maltezos was marketing the long side of the Timberwolf CDSs to Basis and Goldman's own internal estimate of what they were worth. However, one party would benefit from this price fall. Unbeknownst to Basis, 36 percent of the short interest on the Timberwolf CDSs was held by a single counterparty—Goldman. As the CDO lost value, Goldman made money.[26]

Again the Levin Committee report's e-mail trail provides chapter and verse on how Goldman marked down the value of the Timberwolf securities but did not pass on the information to its clients. On May 11, 2007, Sparks e-mailed Richard Ruzika about the problem of unsold CDO securities in Goldman's factory/warehouse, including Timberwolf: "I posted senior guys that I felt there is a real issue. . . . [W]e are going to have a very large mark . . . multiple hundreds. Not good." On the same day, David Lehman of the Mortgage Department announced that Goldman was going to undertake a detailed valuation of the CDO securities still in its factory/warehouse, including again Timberwolf.[27]

When on May 20, 2007, the department presented its findings to copresident Gary Cohn and other senior managers, it reported that it was "most concerned about the CDO position, comprised of the recent Timberwolf and Point Pleasant transactions. The lack of liquidity in this space and the complexity of the product make these extremely difficult to value." In fact, in the preceding days, the Mortgage Department's analysts had come up with some alarming estimates of Timberwolf's loss of value, dragged down by the collapse of the residential mortgage-backed securities included in the CDO. One analyst reported that "based on a small sample of single A CDOs for which we have a complete underlier marks, we believe that the risks of the RMBS underliers are frequently not fully reflected in the marks on the CDOs. . . . [I]f that were the case the price of the A23 tranche of Timberwolf would actually be 35–41 cents on the dollar, depending on the correlation." A week later the same analyst lowered his price estimate to 24 cents on the dollar.[28]

However, these internal estimates of Timberwolf's loss of value were not reflected in the prices Goldman charged its clients, including Basis. In a May 14 e-mail to Thomas Montag, Sparks explained his Timberwolf pricing strategy: "I think we should take the write-down, but market at much higher levels. I'm a little concerned we are overly negative and ahead of the market, and that we could end up leaving some money on the table—but I'm not saying that we shouldn't find and hit some bids." One senior Goldman executive, Harvey Schwartz, warned Sparks and Montag that there was an ethical dimension to their proposed pricing strategy: "Don't think we can trade this with our clients and then mark them down dramatically the next day. . . . [N]eeds to be a discussion if that risk exists."[29]

Schwartz's advice was ignored but was then vindicated literally within days of Goldman selling the Timberwolf securities to Basis for what soon were shown to be grossly inflated prices. Even though on June 13 David Lehman of the Mortgage Department had told Thomas Montag that Goldman's internal valuation for the Timberwolf AA securities it was selling to Basis was $65, on June 18 the sale went through at $84 for securities rated AAA and $75 for securities rated AA. Just two weeks later, Goldman went public with its hitherto internal prices, informing Basis that the Timberwolf AAA and AA securities had lost value—it did not say by how much—and required the hedge fund to post additional collateral of $5 million against money Goldman had loaned Basis to make the purchases.[30]

Thereafter, Timberwolf prices fell very rapid to levels consistent with Goldman's internal valuation: falling to $65 and $60 for the AAA- and AA-rated securities on July 12 and to $55 and $45 on July 16. As prices collapsed, Goldman's demands to Basis for more collateral came thick and fast: for $5.1 million on July 11, $8.1 million on July 12, $12.4 million on July 16, and $5.1 million on July 17. By the end of July, Basis liquidated the hedge fund, and Goldman bought back the AAA- and AA-rated Timberwolf securities for $30 and $25. Meanwhile, as counterparty on the short side with 36 percent of the credit default swaps issued by Basis and other Goldman clients, Goldman cashed in as its insurance policies came good. Throughout July Basis asked Goldman for data that would justify the support of its downward adjustment of Timberwolf securities. According to a complaint filed against Goldman in the New York courts by Basis in October 2011, Goldman consistently refused.[31]

Basis had filed this complaint against Goldman for "knowingly making materially false statements with the sale of Timberwolf" and Point Pleasant, another CDO. As of November 2013 the case has yet to be decided by the court.

THE ROLE OF the investment banks in the financial crisis of 2007–2008, including the role of Goldman Sachs examined here, is the most disturbing recent example of the economic and the unethical coming together with catastrophic consequences for the former. The trading teams at Goldman and the executives who presided over them were inhabiting an extreme variant of a closed world, where their reality was populated by the electronic images and databases of their systems. Their obsessive focus on these images, driven by an equal obsession for the bottom line, excluded the human realities tethered to the symbols.

So they cast a blind eye to the interests of the householders being pumped for subprime mortgages; the unscrupulousness of the mort-gage brokers lining up the subprime borrowers; the duplicity of the ratings agencies, handsomely rewarded, in transforming mortgage dross to AAA gold; and the interests of the bankers' clients who, at the bankers' persuasion and deceived by them, were the end pur-chasers of the flawed derivatives. The cumulative impact of this fu-sion of technology, greed, and moral blindness, duplicated from one end of Wall Street to the other, was global economic meltdown.

So far only a single Goldman employee, the hapless Frenchman Fabrice Tourre, has been held to account for what Goldman did during those years. In a lawsuit brought by the SEC, Tourre was convicted of civil fraud by a New York City jury in August 2013.[32] But as a midlevel processor and salesman of derivatives, Tourre was a very small cog in the Goldman machine, a conveyor belt for or-ders coming down from above. When will the real culprits be held to account?

6

EMOTIONAL LABOR

ALL THE VARIANTS OF COMPUTER BUSINESS SYSTEMS LOOKED AT so far have had information technology as their component: IT fused with machines in manufacturing, IT fused with computers as machines in services. The management of "emotional labor" in human resource management is a case study of how the philosophy of CBSs with its emphasis on *process* and *process reengineering* can show up in places where information technology itself is absent. In its place, HRM experts rely on a model of the human mind as itself a process machine whose inner workings can be modified in the interests of corporate efficiency. This raises the haunting question of whether there are any aspects of our lives public or private that are beyond the reach of *process*.

The concept of "emotional labor" has its origins in Arlie Russell Hochschild's seminal work *The Managed Heart: The Commercialization of Human Feeling* (1983). Hochschild took the airline stewardess as her paradigm emotional laborer, but the practice encompasses tens of millions of middle- and lower-income service workers—shop, hotel, and restaurant workers; secretaries and receptionists; nurses and home-care workers; and of course the workforce of customer relations in its entirety. Hochschild defines their labor as one that "requires [the employee] to induce or suppress feeling in order to sustain the outward countenance that produces the proper state of mind in others."[1] Whether at the Walmart checkout or the first-class cabin of a commercial airliner, employees subject to emotional labor must promote the feel-good factor among their customers.

Such emotional labor draws heavily on our reserves of enthusiasm and empathy, sources of self that, in Hochschild's words, "we honor as deep and integral to our individuality."[2] As a progressive humanist, Hochschild's concern is that these demands corrode and distort this inner self and blur the distinction between the self that is "integral to our individuality" and the self we have to create in order to fulfill our obligations as emotional laborers. This is the commercialization of feeling that Hochschild refers to in her book title.

The strategies that employees rely on to deal with the demands of emotional labor have their origins in private life, but in the course of their mutation from the private to the commercial sphere, these strategies have to undergo a radical transformation. In the private sphere, our reliance upon them is shaped by us and by the rhythms in our lives. But once transposed to the commercial sphere, the strategies become components of production and the white-collar assembly line, and the demand for them becomes relentless day after day, week after week, month after month. It is in this industrial context that Hochschild is concerned with the self that is "integral to our individuality" and fears for its integrity.

Hochschild's work has helped spawn a vast scholarly literature on emotional labor, but whereas Hochschild's concern is with the integrity of the managed heart itself, most of the scholarly literature is concerned with how the strategies of emotional labor can best contribute to efficiency and the bottom line. Trawling through the uninspiring literature of human resource management, I came upon an article entitled "A Time-Based Perspective on Emotion Regulation in Emotional Labor Performance." It appeared in the HRM periodical *Research in Personnel and Human Resource Management*, which comes out annually in volume form and is a flagship journal of the discipline. This introduced me to the scholarly literature such as "Emotion-Regulation Theory Applied to Emotional Labor," "General Antecedent-Focused Emotion Regulation," "Response-Focused Emotional Regulation," and "General Predictions for Surface and Deep Acting Based on Emotion-Regulation Theory."[3]

If this sounds like a field of scholarship ripe for spoofing in the manner of Alan D. Sokal's hoaxing of social science theorizing in the scholarly journal *Social Text,* it is.[4] The emotional labor field is to the fore in its use of the most constipated language of social and management "science" to describe matters of deep moral and human significance, but also chilling in its provision of "research" that corporate HRM managers can then rely on to achieve a more effective "emotional regulation" of the workplace. To get a sense of how "emotional regulation" theories work out in practice, it is essential to avoid the dispiriting abstraction of the texts and to look closely at how its control systems might work in the context of real life.

Arlie Hochschild provides one in her study of the emotional labor of airline stewardesses in *The Managed Heart.*[5] Faced with deregulation in the late 1970s, the airline companies resorted to what Hochschild calls "speedup," the attempt to compensate for lower ticket prices by squeezing more passengers into their planes, but without increasing the number of cabin staff available to look after them. The speedup then experienced by the cabin staff was identical to the speedup experienced by Amazon workers in the company's "fulfillment centers" and Walmart workers stocking the shelves in the local Sam's Club. With the deterioration of the staff-passenger ratios on the airliners, and with time available for the "processes" of cabin work unchanged, the attendants had to work faster and so with less time available per passenger both for the physical dimensions of their work—serving meals, checking seat belts—and for the psychological side of charming passengers with feats of "emotional labor."

The cabin crews responded to the company's "speedup" with what Hochschild calls a "slowdown," although this could not take the form of performing the physical tasks of cabin work more slowly, as General Motors and Ford workers slowed down the line in the great strikes of the 1930s. On the airliners the passengers still had to be seated, seat belts fastened, and meals served within times dictated by airline flight schedules. The attendant's "slowdown" instead took the form of a refusal to smile with the glowing fulsomeness demanded by the airline company's "display rules." In Hochschild's words, "They smiled less broadly with a quick release and no sparkle in the eyes, thus dimming the company's message to the people. It is a war of smiles."[6]

From a management perspective, this facial slowdown was highly subversive because, if there were airline competitors who had managed to keep their attendants smiling with preslowdown luster, then passengers might start abandoning the smile-defective airline for its more welcoming competitors. The companies exhorted their attendants to "smile more" and "more sincerely" at an increasing number of passengers, but the attendants resisted and the slowdown continued. Here, however, there was a dimension to their resistance that reflected the labor-market conditions of thirty-plus years ago and that no longer hold today. Attendants at American,

Pan American, and United were able to form an independent union, and the union gave them the confidence and the power to resist. Although they did not know it at the time, the attendants had managed—just—to squeeze in under the wire before Ronald Reagan's defeat of the air-traffic controllers in 1981 and an accompanying management offensive against unions that would greatly lengthen the odds against successful workplace organizing of this kind.

But let us engage in some counterfactual speculation and suppose that the attendants' slowdown was taking place in today's labor markets and not those thirty years ago, and so with a weakened attendants' union or no union at all and with the attendants highly vulnerable to management counterattack. Let us further assume that we are HRM specialists brought in by senior management to end the slowdown and restore the processes of "emotional labor" to the levels mandated by the company's "display rules." How would we go about doing this? It is here that the whole superstructure of research and theorizing created by the emotional labor theorists of the HRM world comes into its own and points the way forward.

In this scholarly literature, it is significant and revealing that the word *process* features very prominently in descriptions of the mental and emotional phenomena that the HRM operators need to work with. The reliance of emotional labor theorists on the language of "process" allows them to make their work comfortable and familiar to corporate HRM operatives steeped in the disciplines and mind-set of process, and they are the theorists' most significant audience. With process to the fore, the HRM operatives can think of their work as a kind of *Business Process Reengineering* of the soul, whereby, in order to tidy up and streamline the processes for reengineering purpose, they must first map them in their unreformed, lapsed state—exactly as prescribed in reengineering textbooks.

In a scholarly survey of leading "emotional management" theories, esteemed within the subdiscipline, Professor Alicia Grandey of Pennsylvania State University provides chapter and verse on the leading role of "process" as an analytical tool of emotion-management theory. She draws on the work of J. J. Gross of Stanford, a behavioral psychologist who seems to have been a kind of guru of emotion theory for the subdiscipline. Gross defines the subject matter of emotional regulation theory as "the processes by which individuals influence which emotions they have, when they have them, and how they experience and express these emotions." Gross proposes "an input-output model," whereby "individuals receive stimulation from the situation and respond with emotions."[7] In plain English: we respond emotionally to events happening around us and communicate this response to others in the social environment.

But what happens if, viewed from an HRM perspective, the input-output model starts breaking down, as it did in the airline cabins of thirty years ago with the outbreak of the "war of smiles" and with the stewardesses resisting management's "display rules" for emotional labor, substituting instead their quick-release smiles and their nonsparkling eyes? In her account of the remedial tools available to HRM operatives, Grandey again cites Gross as the originator of the two analytic concepts that dominate HRM thinking about the regulation of emotional labor: "antecedent-focused emotion regulation" and "response-focused emotion regulation."[8] At this point it is again essential to avoid the abstraction of the scholarly texts and to elaborate the meaning of these two variants of "emotional regulation" in the context of real life. Because these two forms of acting have their origins in the private and not the public sphere, it is with the former that the opening definitions should be focused. Once that is done, it then becomes much easier to see how they mutate when reproduced in an industrialized workplace setting.

Foremost among the rituals of private life that can be compared to the workplace in their demands for "emotional labor" is the wedding ceremony. Guests at weddings become subject to informal but nevertheless strong "display rules," or social conventions, which for some guests may require recourse to strategies of "emotional labor," as elaborated by J. J. Gross and the emotional labor theorists. Ideally, attendance at a wedding should require no recourse to emotional labor at all. The wedding celebrates the love of the bride and groom for one another and their future happiness together. Guests who love and admire the wedding pair express joy and happiness as a natural outgrowth of their positive feelings for the betrothed.

But there may be a minority of guests who entertain less wholesome sentiments and are present at the wedding only at the strong behest of friends and family members, and it is with these potentially troublesome guests that the rather Scrooge-like theorists of emotional labor are most concerned. One such guest might be an ex-boyfriend of the bride spurned by her for another man—the bridegroom. This guest may now think of the bridegroom as a cad and a bounder, the bride as a hussy, and their friends and relatives as beneath contempt. So once the ceremony is over and this dissident guest arrives at the reception with its babble of celebration and high spirits, how should he behave?

His natural inclination may be to express his feelings of anger and resentment by mooching around the reception, exuding discontent and bad-mouthing the bride and groom. But this would constitute a very serious infraction of the display rules governing behavior at weddings, risking in polite society a disastrous loss of reputation and eventual social ostracism. The guest may therefore feel the need to act in defiance of his inner feelings and somehow reproduce in his outward behavior the joy and the high spirits swirling all around him. Here J. J. Gross and the emotional labor theorists have laid out the alternatives open to him, in their singular language. He may just decide to fake it or, in the language of the theorists, have recourse to "response-focused emotion regulation." This is the first of the two acting strategies that feature prominently in the scholarly literature.

In Gross's words, this "response modulation occurs late in the emotion generative process after response tendencies have been initiated." Again in plain English: the aggrieved wedding guest represses his inner feelings of anger and resentment already present within him as part of the "emotion generative process."[9] He represses also their outward manifestation in frowns and bad-mouthing and acts out as best he can the role of someone in high spirits, hoping that he will convince the assembled company and avoid becoming an object of gossip. What the wedding guest does not do with this "response-focused emotional regulation" is to try to induce within himself the positive emotions appropriate to the outward behaviors that he is acting out, which may make them seem more authentic. So there is a brittleness to this acting that makes it vulnerable to what the theorists call "leakage"—an unwanted eruption of the repressed emotion that may rise to the behavioral surface and disturb the even flow of contrived jollity.[10]

This more exacting task of inducing within the subject the emotions appropriate to the behavior demanded by display rules, whether public or private, is the second of the two acting strategies and a central concern of emotion theorists under the rubric of "antecedent-focused emotional regulation." Gross has listed the four strategies that individuals can follow in adjusting their emotions:

situation selection, situation modification, attention deployment, and cognitive change.[11] The wedding guest engages in "situation selection" if, when he arrives at the wedding reception from the church, he takes one look at the assembled company and goes home. With "situation modification," the guest goes to the reception, sits down with a bottle of wine in a remote corner of the tent, avoids the need for enforced jollity, and hopes that no one will notice.

These two "antecedent-focused" strategies are more available to wedding guests than to employees working at call centers or Walmart stores. It is the last two of Gross's "antecedent-focused strategies," attention deployment and cognitive change, that are forcefully present in both the public and the private spheres. With "attention deployment," the wedding guest does not leave the reception or retreat to a distant corner of the tent. He remains with the celebrants, but he focuses his attention not on the dispiriting scene that surrounds him, but on memories of a past luminous wedding that enthralled him, and he does this in the hope that the positive emotions associated with this memory will rise to the surface, lifting his spirits and making him authentically joyous—but authentic in the context of his past memories and not of his experience of the present.

With "cognitive change," the guest does not take refuge in such past memories but alters his perceptions of the living present in ways that can evoke the positive emotions bound up with these changed perceptions. He might begin looking at his fellow guests as objects of mirth, not of resentment. He sees a couple engaged in a passionate affair acting as if they were complete strangers; he sees another couple about to embark on a ferocious divorce gazing at one another as new lovers. He sees young academics fawning over a grizzled professor who despises them. He sees the bridegroom and remembers his dubious claims of gentility and his bounced checks at the club. He marvels at the human comedy and finds himself beaming and even laughing out loud, and his neighbors at the reception look at him and whisper to one another about how magnanimous he is as the great love of his life is borne away by another man.

However, emotional labor theorists are more concerned with "attention deployment" and "cognitive change" in the context of the workplace than at wedding receptions. What form do the two leading processes of emotional labor take in the workplace? For "attention deployment," Hochschild cites the example of a young women working in a department store who whistled opera arias to herself so that fond memories and feelings about *La Traviata* and *La Bohème* replaced as the focus of her attention the rasping demands of irate customers. For cognitive change, she cites an airline company that encouraged its cabin crews to think of passengers as wayward children to be indulged with the warm sentiments of motherhood.[12]

Although there is a formal, structural similarity between the performance of the "processes" of emotional labor in the contexts of public and private life, there are also two characteristics of their performance in the workplace that are not present in the private and radically change their nature. The first difference is that for the wedding guest, unless he suffers from social nymphomania, the need to perform this elaborate and stressful feat of emotional labor arises infrequently and at his own discretion. He does not have to keep going to wedding receptions or their equivalent.

But employees subject to corporate display rules and the emotional labor attached to them must, as we have seen, repeat these processes of psychic mutation day after day, week after week. Also, unlike wedding guests, employees may not have the freedom to choose which form of emotional labor they might pursue. "Situation change" and "situation modification" are effectively ruled out. The choice is between "response modification"—repressing felt emotions and faking false ones—and "antecedent-focused regulation," with its "attention (re)focus" and cognitive change. But these choices fall within the remit of HRM, and a leading task of emotional labor theorists is to work out which of these variants of emotional labor might be best for employee productivity and the bottom line.

In the research literature, there is a bias in favor of "antecedent-focused regulation," on the grounds that employees forced to engage in its opposite, "response-focused regulation," over prolonged periods are more likely to suffer from emotional burnout. Indeed, Grandey is describing a scholarly "burnout literature" concerned specifically with this problem.[13] Grandey defines *burnout* as a condition arising when "a situation induces repeated emotional responses that the employee must regulate," with the employee having "little in the way to replenish those emotional resources being spent." So the employee may experience the symptoms of "emotional exhaustion . . . energy depletion and fatigue"—burnout.

We have already come across the concept of "leakage" as a warning symptom of burnout, those unwanted eruptions of repressed emotion that come to the surface if response modification is practiced for too long. Grandey is also describing something much worse that is the emotional labor equivalent of a nervous breakdown. Here, with burnout, the employee's underlying distress takes over completely and pushes aside the surface contrivances of "response modulation." The job environment "may induce an emotion response in the employee—anger, sadness, anxiety—and behaviors may follow that would be inappropriate for the encounter—verbal attack, crying, complaining."

Why should antecedent-focused response modulation be any less vulnerable to the perils of employee burnout than its response-focused opposite? The answer is that because the emotions induced by the first processes are, at least at the time they are felt, genuine, their expression in outward behavior flows naturally from their existence and so is not faked and is less susceptible to burnout. According to Gross, this "deep acting" convinces employees that they really feel the way they are trying to express. Although this process is still effortful, "it may lead to an expression that is perceived as more genuine than when an employee surface acts."[14]

Also there may be management bias in favor of the emotional reappraisal or "self-talk" because the willingness of employees to undergo it shows that they have "good faith" toward the organization and are prepared to undergo emotional reengineering in its interests. However, others have pointed out that antecedent focus is not risk free, and these have of course included Hochschild. Those engaged in "attention displacement" are in effect daydreaming and may therefore be vulnerable to workplace harassment from colleagues and supervisors.[15] The young store worker whistling opera arias may be pushed around by supervisors or coworkers who don't like her choice of whistling and think, with reason, that she is not focusing on her work and so may be vulnerable to workplace harassment from coworkers.

Similarly, employees practicing "cognitive change" may fail to defend themselves against threatening and bullying behavior from customers. Airline cabin attendants trained to regard passengers as wayward children may fail to notice that their nemesis in first class is not a wayward child but a high-level attorney or corporate executive quite capable of the unchildlike behavior of firing off a blistering letter of complaint about their work to the airline CEO. All these criticisms point to what is a glaring defect of both techniques of cognitive change: both prevent employees from focusing their full attention on the task they have to perform as it really is, one by encouraging them to take refuge in memory and fantasy, the other by surrounding their work with perceptions that may be delusional.

Finally, there is Hochschild's criticism of this emotional re-engineering that goes beyond comparisons of the relative utility to business of the two variants of cognitive regulation. She gets to the heart of the matter when she asks, what right do businesses have to manipulate and market these "sources of self that we honor as deep and central to our individuality," and why shouldn't we describe this manipulation simply as brainwashing? At its worst, emotional labor seizes upon the feelings, perceptions, and judgments that constitute who we are, pushes them aside, and substitutes a body of imposed perceptions and feeling that originates not in ourselves but in the judgment of HRM specialists relying on research telling them which mental mix will best serve the interests of the business.

THERE IS PERHAPS an exit from the grimly dystopic world of emotional labor and one that does not sacrifice the quality of services provided by employees. In an article that appeared nearly thirty years ago in the *Journal of Marketing,* three academic specialists in marketing, A. Parasuraman, Valarie Zeithaml, and Leonard Berry, pointed to a possible means of escape. They included in their texts the result of a comprehensive survey that asked focus groups of consumers to describe what for them constituted quality in services. What is striking about the survey is the respondents' lack of concern with the "products" of emotional labor as a determinant of service quality and their overwhelming focus on the competence of service workers in doing their jobs, which includes knowledge of their work, skill in explaining it to customers, an understanding of the customer's needs, and the ability to respond to these needs and so provide individualized service for the customer. "Emotional labor" is present in the survey under the heading of "courtesy" and includes politeness, respect, consideration, and friendliness.[16]

The marks of service quality chosen by the respondents to Parasuraman, Zeithaml, and Berry's survey correspond almost exactly to the service provided by the two Microsoft customer service agents whose work I described in chapter 3. They were highly knowledgeable about Microsoft's XP operating system and quick to respond to the customer's needs in language that could be readily understood. Under "courtesy" the two agents were certainly polite, respectful, and considerate, and I would add a certain buoyancy stemming not from their "emotional labor," but from their confidence in their skills and their knowledge that they were doing their job well. But does that amount to "friendliness," and was there an emotional dimension to their work at all?

This gets us deep into philosophical territory, but we have already been there with the emotion theorists' reliance on Cartesian mind-body interaction, with inner emotions manifesting their presence in outward behavior. In 1949 the Oxford philosopher Gilbert Ryle wrote *The Concept of Mind*, among the most influential philosophical works of the twentieth century.[17] In it Ryle sets out to show that much of the language we use to describe and characterize outward behavior does not entail any reference to inner emotional states or indeed to inner states of any kind. At times Ryle seems to go further, by denying that language *ever* entails such references or denying that such inner states even exist—evidence of how philosophers can get carried away by their arguments and deny truths evident to every sentient being.

We know that inner states exist because we experience them every day. But Ryle's central insight into what he calls the language of dispositions is compelling and can illuminate the language of service excellence as it features in Parasuraman, Zeithaml, and Berry's survey: responsiveness, competence, understanding. The presence or absence of these qualities in service work depends not on the presence or absence of accompanying emotion, but simply on how work is performed over time, whether problems are or are not resolved, and whether the work is done in a manner that suggests competence and confidence.

This can be no less true of the candidates for emotional labor in the Parasuraman survey, such as courtesy, politeness, respect, and consideration. These qualities again are not present as separate sound tracks of the mind running parallel to their outward behavioral displays. They are embedded within the behaviors themselves as everyday expressions of respect for our fellow citizens, no less genuine for being present without conscious effort or accompanying emotion. They form part of our public selves shaped in countless transactions with fellow citizens and take place beyond the workplace as well as within it. They are rituals of civic life that, even though performed instinctively, have an ethical value that we would strongly defend if challenged, for without such rituals, everyday life becomes a minefield of antagonisms and misunderstandings.

The great virtue of such displays is that, as expressions of our public selves, they do not oblige us to reach down into our private selves and induce, repress, or modify our feelings as reengineers of emotional labor. One might then ask why the HRM theorists and practitioners do not simply abandon the whole deeply dysfunctional enterprise of emotional labor and simply build their code of employee conduct on the solid foundation of the employees' public, civic selves. But this cannot be done just be pushing aside one set of behaviors and substituting another, like changing the reel in a movie projector.

In the workplace the public, civic self of employees has to be sustained not just by the employee's own civic life beyond the workplace, but also by the workplace equivalent of a civic life. For employees consistently to act with high competence and courtesy, a support system has to be present, but one that is increasingly absent from the American workplace, especially for middle- and lower-income employees. The support system should include good education and training, time and autonomy to do work well, job security and financial rewards if the job is done well, and with an independent voice for employees in their dealings with management.

In a piece entitled "Social Legitimacy of the HRM Profession," included in *The Oxford Handbook of Human Resource Management,* MIT economist Thomas Kochan has a section titled "Breakdown in the (American) Social Contract."[18] Kochan describes himself as a "card-carrying member of the Society for Human Resource Management and of the National Academy of Human Resources"—leading associations of the HRM profession. He is also an economist of strong progressive leanings, so his presence in the HRM world makes for an unusual perspective on workplace issues. In his piece Kochan lists some of the workplace developments that over the past twenty years have killed the old social contract binding employers and made the creation of any new contract extremely unlikely.

The killers of the contract include increased working hours for individuals and family units; increased inequality of income and stagnant or declining real wages for a majority of the workforce; the break in the historical relationship between profits, productivity, and real wage growth; loss of retirement income and shifts in the pension risk to employees; declining health care coverage and shifts of cost increases to employees; loss of employee voice at work as labor-movement members decline to pre-1930 levels; and increased use of layoffs not as a last resort but as a routine aspect of corporate restructuring. To the list should be added the increased pace of work dictated by CBSs, its intensive targeting and motoring by "performance evaluation" systems, and its deskilling of employees with expert systems.

This multifaceted deterioration in the condition of labor helps explain why the theorists and practitioners of emotional labor devote so much effort to the repression of negative, work-disrupting emotions before trying to replace them with more customer-friendly ones, which they think will be good for sales. But looking at this project with even a modicum of historical perspective, it seems doomed. The harsh, unforgiving workplace described by Kochan yields a negative emotional labor all its own, and the longer this workplace endures, the more entrenched becomes the emotional armor that employees must rely on to protect themselves against a hostile world, thus cynicism, resentment, emotional withdrawal, and the withholding of loyalty from employers who show no loyalty to them.

With this negative emotional geography in place, the emotional labor of HRM becomes a desperately affixed bandage with the near-hopeless task of neutralizing this emotional malaise and replacing it with something more positive. But this much-desired regime of positive emotions cannot be imposed by fiat and in defiance of reality. Sooner rather than later, reality will prevail, and the whole contorted structure of "emotional labor" will collapse, with the "anger, sadness, and anxiety" of Grandey's burned-out workplace setting the tone for much of the service economy. In fact, there already exists a developed industrial economy where this collapse has reached an advanced stage, and of all the developed economies it is, with the exception of Canada, the one nearest to the United States in history, culture, and political economy, namely, the United Kingdom.

As a UK citizen who has spent half his adult life in that country, I have watched this collapse gathering pace, especially during the thirty-year life of the neoliberal economy bequeathed to the United Kingdom by Margaret Thatcher from the 1980s onward. The British case is a dire warning to the United States of what will happen to it if it allows its present downward drift to continue. The UK service workforce shares many characteristics with its US counterparts: poor secondary education, even poorer vocational education, and a similar workplace bias in favor of management that yields stagnant real wages, growing inequalities of income and wealth, weak unions, declining benefits, and frequent recourse to outsourcing and layoffs. However, there are two negative characteristics of the UK service workforce that the US workforce does not yet share to the same degree, but will surely do so if present trends continue. The first is the existence of deep and antagonistic differences of class culture that in Britain can make the daily exchanges of the service economy highly problematic, especially in London and the South of England, the region now overwhelmingly dominant in the United Kingdom both economically and culturally.

This is also the region of the UK where the refinements of middle-class gentility run deepest and where they come up against the remnants of an increasingly antagonistic working-class culture undermined by the near disappearance of the British manufacturing economy and the failure of successive UK governments to provide the education and training needed for employees to succeed in a service-dominated economy. In such an economy the transactions of the service sector are increasingly dominated by the defensiveness of employees who feel disadvantaged by poor education, a lack of proper training for the job, lack of opportunities for advancement, and a culture increasingly despised by the middle class—though in a typically British way, this contempt is only hinted at.

Attempts to cultivate emotional labor in such unpromising soil have rarely succeeded, and British shops, offices, call centers, banks, hotels, and railway stations are not welcoming places. Visitors to Oxford from the employee-friendly northern European economies of Scandinavia, Germany, and the Netherlands have sometimes asked me whether Britain is in a state of incipient revolution, such are the currents of antagonism running close to the surface of the British service workplace. My answer is that the weight of conformity pressing down from above precludes this, as does a history of hierarchy and stratification stretching back undisturbed to the seventeenth century and beyond. However, the North London riots of 2011, when gangs of the unemployed and the low paid took to the streets and looted stores and supermarkets, are signs that the control mechanisms of history and class may be wearing thin.

A second source of British dysfunction derives from the persistence of these economic and social failures over time. As the generations of the disadvantaged succeed one another, each succeeding generation becomes further removed in time from a period when there was a preceding generation gainfully educated and employed. For a growing core of the deprived, this backward time travel has to extend all the way to the 1950s, when Britain was still a major manufacturing power and when a majority of the British adult male population had been subject to military service in the two world wars and the Cold War. With the passage of time, much of this core is becoming virtually unemployable, such is their lack of language and social skills, so when they are employed, their impact on the quality of service is severely negative.

Both in the United States and in Britain a contrarian indicator of this growing industrialization of the service economy is the rapid and parallel growth of the concierge economy, in which very high-income consumers use their financial muscle to escape reliance on the defective, mass-produced services available to middle- and lower-income consumers. So there are concierge doctors on Park Avenue and mass-production doctors with HMOs, concierge personal bankers at the Goldman Sachs private bank and mass-production bankers (if you can reach them) at Citicorp. In the concierge economy, the relationship between technology and work is turned on its head, and information systems are used to supplement rather than replace the skills of employees. There are no digital scripts at the Goldman Sachs private bank.

The UK again provides the most spectacular example of a concierge service economy existing alongside a "standard" economy as a visible monument to inequality. Concierge London is a narrow strip of real estate stretching from the financial district, the City of London, at its eastern extremity to the hotels, casinos, and hedge-fund boutiques of Mayfair at its western. This strip has the feel of an offshore economy stranded onshore, an inflated Monaco that somehow finds itself surrounded by a local economy whose most blighted districts, including those nearby in inner London itself, increasingly resemble the Italian Mezzogiorno in their decline and hopelessness. Nonetheless, this narrow strip is a big revenue earner that helps prevent the United Kingdom from falling in the European rankings to somewhere between Italy and Spain.

Concierge London is a creation of footloose capital both corporate and personal, much of it driven by tax avoidance, and by the financial speculation that brought the global economy to its knees in 2007–2008. Such origins account for the kitschy, brittle feel of concierge London, especially as encountered in the house magazines of the *quartier,* the *Financial Times' How to Spend It* supplement and the local London edition of *Vanity Fair.* The dominant players of concierge London have been the global financial conglomerates, Goldman Sachs, JPMorgan Chase, Citicorp, Deutsche Bank, and UBS, relying heavily on London as a trading center during their flush years, as well as a globally footloose plutocracy parking their money in London real estate as a safe haven, notably Russian oligarchs, Middle Eastern sheikhs and despots, American speculators for whom even the SEC regime on Wall Street is too intrusive, and more recently anxious Europeans—French, Italian, Spanish— fleeing the perils of the eurozone.

More and more, the institutions of the British state—monarchy, government, Parliament, bureaucracy—have the look and feel of a respectable but ineffective Potemkin facade, behind which the real business of financial and real estate speculation in concierge London can be carried on. In such a distorted economy, the class stratification that so stunts British society actually comes into its own. British gentility, with its ancient aristocratic and military origins but increasingly diffused among the middle classes in the course of the twentieth century, has become a highly marketable form of "emotional labor" much in demand in concierge London and indeed beyond. No Russian oligarch coming to London wants to put up with British middlemen and -women speaking the faux proletarianism of "estuary English," much in fashion during the 1990s and once favored by Tony Blair.

Moving within this London concierge economy, it sometimes

seems as though much of the British elite have themselves become high-end concierges because, with the core of wealth and power coming from the outside, the local British who deal with the core must assume a certain concierge deference, and this includes not only genteel young women working in upmarket art auction houses or real estate agencies, but also lawyers, accountants, bankers, management consultants, public relations operatives, and boutique asset managers in their Mayfair town houses. It is ironic that many among this concierge elite are descendants of those who, at the height of the imperial era a century ago, expected a similar deference from "foreigners," especially those from the colonies. Now the boot is on the other foot, with a vengeance.

Looked at functionally, the London concierge economy and its New York equivalent work. Their high-income clients demand good service, and on the whole they get it. Companies providing concierge service pay employees well, make sure they are properly trained for the job, make sure that they look good and are well dressed, and provide time for the job to be done properly. In London the upper-middle-class ethos that predominates on the British side of the concierge economy copes well with the problems of "emotional labor." With its strong imperial and military roots, it provide a version of the public self that, though sometimes verging on bossiness, is confident, efficient, courteous, and with the required hint of deference. These "display rules" do not require expressions of private emotion. The elaborate stratifications of English social life provide an equally elaborate public self that can be deployed in the workplace without recourse to private displays.

Concierge economies show that business can provide good service provided that there are high-end clients with the power and wealth to insist on it. Concierge economies can even come up with a solution to the problem of "emotional labor." But these gilded ghettos remain flawed and highly visible symbols of a decaying polity as long as they are surrounded by "standard" economies where underpaid and poorly trained employees provide bad service for the great majority of consumers. Is it beyond the power of business to provide good service not just for the high-income few but also for the middle- and lower-income many, when "good service" means not just the efficient performance of tasks but also an accompanying human dimension that avoids the demeaning psychic contortions of "emotional labor"?

Some of the best examples of how good service can be available to everyone, irrespective of income, are to be found in the northern European economies of the Netherlands, Germany, and the Scandinavian countries. With the deepening of the eurozone crisis, these economies have in many American eyes become lumped together with the economies of the European South—Italy, Spain, and Greece—as case histories of profligacy and mismanagement. But this amalgamation obscures the achievement of the northern Europeans in often finding solutions to core problems of the "nonconcierge" service economy that have so far eluded the Anglo-Americans. Their record shows that one essential ingredient of success is an effective system of vocational education that provides the nonacademic with the skills needed for their work, as well as a foundation for the continued development of these skills as careers progress. In the United States, President Obama has, as we have seen, talked of upgrading the community colleges for this role.[19] Judged by its ability to keep youth working during the eurozone recession, the German system of vocational education, divided between technical schools and the workplace, is the European leader. While youth unemployment in France, Italy, and Spain was, at the end of 2012, at 16, 23, and 50 percent, respectively, in Germany it was a modest 7 percent.[20]

The German system has its origins in manufacturing and provides the foundation for German global supremacy in high-end engineering, as we saw in the case of the machinists at the Treuhand plant in Chemnitz. But successive German governments have extended this system from manufacturing to services so that young Germans taking vocational courses in retail, financial services, or the hotel trade learn not only the routine tasks of the trade as in the United States and the United Kingdom—how to check in a customer and how to prepare her final bill—but also about the structure of the industry and the tasks of management. This knowledge provides the foundation for future advancement. For a UK citizen, it is disconcerting to visit even middle-ranking hotels in Hamburg or Munich and deal with young men and women behind the counter who in the United Kingdom would have the skills and confidence to be middle managers.

However, successful service economies require not only skilled employees trained to do the job well, but also a means of communication between employees and customers that avoids the contortions of emotional labor and the equal stress of transactions that become denuded of civility when emotional labor fails. Here the Scandinavian economies demonstrate the practical value of equality in a service economy. Equality has an economic and social dimension, providing employees with the dignity, confidence, and skill to provide good service for all, irrespective of income and rank. The economic preconditions for this are generous pay and benefits and an equitable distribution of workplace power between management and labor, provided in turn by high rates of unionization. But equally important is a strong tradition of civic equality that gives employees and customers a shared language of courtesy and respect, so that their public selves can manage the transactions of everyday commerce without recourse to emotional labor.

7

THE MILITARY HALF

COMPUTER BUSINESS SYSTEMS HAVE A HISTORY GOING BACK AT least seventy years to the Second World War, and the length and depth of this history have been powerful forces shaping today's systems. Among the most remarkable pieces of evidence we have of this history's role is an article, "Management in the 1980s," which appeared in the November–December 1958 edition of the *Harvard Business Review*. The authors, Harold J. Leavitt and Thomas L. Whisler, were at the time professors of business administration at the University of Chicago. In their piece Leavitt and Whisler are credited with the first-ever use of the term *information technology*, and this alone gives their piece landmark status.

With their predictions about the future of information technology and corporate management, Leavitt and Whisler were not flying entirely blind. When they did their research in the late 1950s, there were already rudimentary civilian CBSs in existence that contained enough of the DNA of future systems for discerning observers like Leavitt and Whisler to figure out how the systems might evolve in the coming decades. In their piece Leavitt and Whisler describe how, at an unnamed manufacturing plant, computer programmers "have had some successes in displacing the judgment and experience of production schedulers," thereby "displacing the weekly scheduling meeting of production, sales and supply people." Such programs were also being worked out in increasing numbers "to yield decisions about product mixes, warehousing, capital budgeting and so forth."

As their title suggests, Leavitt and Whisler were interested not just in describing the systems taking shape around them in their own times, but also in figuring out how the systems might evolve in the coming decades and how they might transform the structure of corporate management. Their forecasts of fifty-five years ago have turned out to be uncannily accurate, more accurate indeed than many of the pieces dealing with contemporary management systems now appearing in the contemporary scholarly literature. Writing at a time when the United States was still indisputably the world's leading industrial power, Leavitt and Whisler assumed that the American industrial model that had taken shape during the first half of the twentieth century would continue to shape how the newly emerging technologies would be used during the century's second half. This is essentially what has happened, and Leavitt and Whisler's forecasts have withstood the test of time exceptionally well.

Leavitt and Whisler's piece is significant as much for its forecasts of managerial winners and losers as it is for their forecasts of what future technologies might look like. They predicted that middle managers would be the big losers. They would lose skills, function, and power; be reduced in numbers; and be paid less. "There will be many fewer middle managers, and most of those who remain are likely to be routine technicians rather than thinkers."[1] The middle loses out because skilled tasks that they had hitherto performed, such as the gathering and analysis of data and the scheduling of production, would be taken over by information systems. These data would also be available to senior management in real time and would open middle managers to the kind of intrusive monitoring from above that middle managers themselves had exercised over their front-line subordinates.

Leavitt and Whisler predicted that IT systems would bring about a centralization of power in the hands of top management and that "the line separating the top from the middle of the organization would be drawn more clearly than ever, much like the line drawn in the last few decades between hourly workers and first-line supervisors." "By permitting more information to be organized more simply and processed more rapidly," information technology "will allow the top level of management to categorize, digest and act on a wider range of problems. . . . [B]y quantifying more information it will extend top management's control over the decision processes of subordinates."[2]

The summit of the organization chart of the future corporation would look "something like a football" resting at the top of the familiar pyramidical hierarchy. Skill, power, and money would move upward to this "little oligarchy of head men" and with a small group of system designers at hand. These were the inhabitants of the football, and the IT specialists among them were the indispensable agents of centralization because it was they who would create the systems that would give top management their panoptic powers. This concentration of power at the summit and the hollowing out of the corporate middle were seen by Leavitt and Whisler as a new variant of Scientific Management, with information technology moving "the boundary between planning and performance upwards." Just as the scientific managers of Frederick W. Taylor's time had taken the planning of work away "from the hourly workers and given [it] to the industrial engineers," so now with the planning and rigorous monitoring of the middle managers' own work, where power would be given to this elite of senior managers and IT specialists.[3]

IN MANY ACCOUNTS of CBS history, including even Leavitt and Whisler's, something vital is missing. These are the military variants of CBSs originating in World War II and the Cold War. If the early pioneers of civilian CBSs in the 1950s and 1960s were resilient in the face of their sometimes dysfunctional systems, it was because they could look back and around, taking heart from the successes of "War IT," of how systems, management, production, information, and weapons had come together on the gigantic scale needed to fight and win the battles of World War II and the theoretical battles of the Cold War.[4] During those decades dominated by war or by fear of war, there was a succession of military-related projects that can illuminate our understanding of the early history of civilian CBSs.

Among these histories were the conversion of the US industrial economy under central government direction from 1942 onward and the waging of technology-intensive campaigns between 1940 and 1970—the air defense of England during the Battle of Britain (1940), the Battle of the Atlantic (1940–1945), the strategic bombing of Germany (1942–1945), and the fighting of the Vietnam War (1965–1975), the first three ending in victory, the third not. This chronology includes two case histories that, although deploying vast industrial and financial resources, depended for their success on the ability of a limited number of elite scientists to surmount major technical obstacles and to achieve specific objectives under a strong pressure of time. These were the Manhattan Project to build the atomic bomb before the Germans did and the Apollo Project to put a man on the moon before the Russians. A seventh project was a hybrid combining the rationality of system wars with the fog of war as conceived by Thucydides, Tolstoy, and more recently Robert Mc-Namara. This was the Allied invasion of France in June 1944, Operation Overlord, whose planning still ranks among the most complex and ambitious operations in human history yet whose managerial rationalism, once the Allied forces crossed the English Channel and stormed the Normandy beaches, quickly gave way to the fog of war.

The final case history is perhaps the most significant: the creation and management of the US strategic nuclear forces from the late 1940s until the end of the Cold War in 1989. These forces comprised by the early 1960s the strategic triad of land-based intercontinental ballistic missiles (ICBMs), long-range bombers carrying nuclear bombs or missiles, and submarines carrying ballistic missiles (SLBMs) and eventually cruise missiles (SLCMs). The management of the US strategic nuclear forces included the construction of the main weapons systems, their maintenance in a high state of readiness so they could survive a Soviet first strike and be in a position to retaliate in a second strike against Soviet military and industrial targets, and the creation of defense and early-warning systems against Soviet bombers or missiles approaching the United States from Arctic Canada.

There were at least six ways in which the ambitions of the CBS pioneers and their corporate bosses in the decades following World War II were inspired and shaped by the achievements of the military and, in the case of the Apollo Project, quasi-military variants of management giantism. This was not simply a question of the achievements of World War II casting a shadow over the postwar decades. This influence was enduring because the management systems of the Cold War, increasingly sustained by information technologies, kept alive and in some ways magnified the influence of wartime system design in the postwar era.

From the 1950s onward, the postwar pioneers of civilian CBSs were working toward "totally integrated management information systems (MIS)," which "promised a new vision of management to a corporate world self-consciously remaking itself around science, high technology, staff experts and systems."[5] This vision began to take shape during the Second World War in campaigns such as the Battle of the Atlantic where multiple technologies were deployed—radar, sonar, wireless, and early-generation computers—that together could provide a total, "holistic" view of the battlefield. The CBS pioneers envisaged systems that could control the operations of multinational corporations with global reach. In World War II the size of the forces subject to centralized command dwarfed the size of the largest corporations existing then or now and covered vast areas: the combined Allied bomber forces in the strategic bombing of Germany ranging over the airspace between central Europe and the United Kingdom; the Allies' armies in the planning for D-day, occupying virtually the whole of southern England; the Battle of the Atlantic, ranging from Greenland to northeastern Brazil; and, during the Cold War, the US strategic nuclear forces deployed all over the world.

The CBS pioneers saw their "total management systems" pivoting on top management exercising central control over their organizations. In World War II information systems were at the disposal of Allied commanders with full powers: General Dwight Eisenhower for D-day; General Carl Spaatz, General Curtis LeMay, and Air Chief Marshal Arthur Harris for the strategic bombing of Germany; Admirals Max Horton and Ernest King for the Battle of the Atlantic;

and Air Chief Marshal Hugh Dowding for the Battle of Britain. Because their subordinates were bound by military discipline, their instructions had the force of orders and were to be obeyed.

For the civilian CBS pioneers, speed of operation and adaptation were needed if the systems were to respond effectively to unexpected events such as failure of a new product, a loss of market share, or the advent of a new technology. In system wars plans had to be constantly changed in light of unexpected developments and setbacks: in the Battle of the Atlantic, the need to adapt to the loss of the German codes in 1942; on D-day, the need to deal with the unexpected reverses on Omaha Beach. In the Battle of Britain, controllers under intense pressure of time had to process a mass of information about incoming German forces in order to provide directions for the defending forces. In the Battle of the Atlantic, air and surface forces had to be deployed under pressure of time in response to intelligence about U-boat movements.

A leading aim of the CBS pioneers in the postwar decades was to automate with computer technologies their "total management information systems." Such technologies began to emerge in strength after the Second World War II from the military side and, from the late 1950s onward, from the National Aeronautics and Space Administration. Although these nascent technologies were also to be found on the civilian side, it was the military side that dominated. The most advanced and ambitious computer projects were bound up with the creation and deployment of the US strategic nuclear forces.

The military side included the SAGE computerized air-defense system deployed to detect and intercept Soviet bombers coming in from the north; the development in the 1950s of the PERT (program evaluation and review technology) system for the management of the Polaris submarine-based missile program; and the duplicated command-and-control systems for US land-based ICBMs that had to be able to survive a Soviet nuclear attack and execute a US second strike in a context of postattack nuclear devastation. The Apollo Program relied on an IBM System 360 Mainframe for communications between Houston ground control and the Apollo spacecraft for the monitoring of the spacecraft's flight path, its condition, the medical condition of the astronauts, and the calculation of the liftoff data to launch the lunar module from the moon's surface.

OF ALL THE BATTLES of World War II, the Battle of the Atlantic was the one in which information technologies deployed on the Allied side got closest to becoming a military CBS, providing a panoptic view of the battlefield and enabling the Allies in the spring of 1943 to defeat the German U-boats with the speed and decisiveness of a maritime blitzkrieg.[6] The Allied technologies included radar, sonar, and early-generation computers that enabled the Allies to break the German codes and chart the movement of the U-boats from their bases in occupied France and Norway into the North Atlantic.

Although the Battle of the Atlantic was fought mostly within a defined geographical space, this space was vast and covered the entire North Atlantic. A map in volume 5 of Winston Churchill's *Second World War* shows the location of Allied shipping losses during what Churchill calls "the crisis of the battle" between August 1942 and May 1943. Most sinkings took place within a space defined to the north by a line linking Newfoundland, the southernmost tip of Greenland, and Iceland and to the south by a line linking the northeast coast of Brazil and the coastline of Sierra Leone, then a British colony on the west coast of Africa.

One of the legends of World War II is that the Allies always had a monopoly of code breaking and could read all German military communications, including those between German naval headquarters in occupied France and the U-boats. But for the first three years of the Battle of the Atlantic, the information war between the two sides was more evenly balanced. From the start, the "visibility factor" strongly favored the U-boats. Allied convoys forty or fifty ships strong, slow moving, giving off smoke, and strung out along the ocean, copiously diffused information about their position, while the U-boats presented minuscule targets in the vast tracts of the Atlantic and were especially hard to find when submerged or moving at night.

The battle of the code breakers, critical to the information side of the battle, sometimes favored the Germans, notably in 1942 and early 1943 when Allied fortunes were at their lowest and the British naval staff conceded that "the U boats came very near to disrupting communications between the New World and the Old." In December 1941 German naval intelligence, the B-Dienst, broke the British Naval Cypher No. 3 used by the US, Canadian, and British navies for controlling the movement of all transatlantic convoys. Then in February 1942 the Germans replaced Enigma with a new naval cypher, Triton, which the Allies could not break until the spring of 1943. For a fourteen-month period, the intelligence war strongly favored the German side, and this was reflected in the statistics for tonnage of Allied ships lost, which rose to 3.7 million between August 1942 and May 1943.

Yet by the spring of 1943, the Allies had effectively won the Battle of the Atlantic. By May 1943 more U-boats were being sunk than Allied cargo ships, with forty U-boats lost in May alone. Allied tonnage lost fell from the 3.7 million tons between August 1942 and May 1943 to just 207,000 between May and September 1943, less than one-tenth the 1942–1943 figure. On May 23, 1943, Admiral Karl Donitz, commander in chief of the German Navy, told Hitler that the Battle of the Atlantic was for the moment lost and ordered the U-boats back to their French bases. How could the fortunes of war have been reversed so quickly, and what role did information systems play in this reversal?

In his history of World War II, Winston Churchill described the Battle of the Atlantic as expressing itself "through statistics, diagrams and curves unknown to the nation, incomprehensible to the public."[7] The Battle of the Atlantic was not just a naval battle but a theater of economic warfare whose goal for the German side was to increase Allied shipping losses to a level where the British economy would lack the oil to support war production and the food to feed the British population. Figures for the volume of British exports and the ratio of ship tonnage loss were therefore vital statistics of the battle. In his history Churchill describes how toward the end of 1940, he became "increasingly concerned about the ominous fall in imports from a rate of 1.2 million tons a month in June 1940 to 750,000 tons in July to 800,000 tons in August." Unless reversed, these were the statistics of defeat.

At the level of operations, the "total information" system that the Allies eventually constructed for the Battle of the Atlantic had to integrate information about the deployment of friendly and enemy forces over the whole vast battlefield with information about the exact location of nearby enemy forces so that Allied warships and aircraft could find and destroy them. It was not until the spring of 1943 that the Allies had both the necessary antisubmarine technologies to plot the exact position of the U-boats as well as the aircraft and warships in sufficient numbers to exploit this information to the full. It was then that the information system for the battle became seamlessly integrated in a way that linked a panoptic view of the gigantic battlefield with a detailed view of the position of a U-boat "wolf pack" off the southern coast of Greenland.

By the spring of 1942, the US and British navies had set up tracking rooms in Washington and London that gathered all the available evidence on the movement and position of the U-boats throughout the northern and southern Atlantic. Cryptologically blinded in 1942 by their failure to break the German T code, the Allies had to rely on a variety of lesser sources to plot the position of the U-boats. Among them were the reports of agents in occupied France and the origin and direction of German radio messages picked up by intercept stations scattered around the Atlantic rim. Even without direct access to communication to and from the U-boats, the Allies were successful in using these lesser intelligence sources to establish the approximate position of the U-boats and to steer the Allied convoys away from them. Between May 1942 and May 1943, 105 out of the 174 convoys that crossed the Atlantic managed to avoid the U-boats. Out of 69 sighted by the U-boats, 23 escaped attack and 30 suffered minor losses. The remaining 16 were heavily damaged.

The coming together of technology and firepower in the spring of 1943 enabled the Allies to change their strategy from one of relying on information about the position of the U-boats to keeping the convoys away from them to a strategy of using this information to lure the U-boats into attacking the convoys and then destroying them with greatly strengthened air and naval forces. The lightning victory of the Allies in May 1943 was the result. The advances in technology that made this possible were the rebreaking of the German Triton code that again provided the Allies with direct access to communications to and from the U-boats, the invention of sea- and airborne antisubmarine radar that enabled long-range aircraft and escort ships to detect the presence of U-boats in day or at night within a radius of five miles, and the jamming of the German Matrix radar that had made it possible for U-boats to detect the presence of approaching Allied aircraft.

By the spring of 1943, the resources available to the Allies had also increased to the point where they had enough long-range bombers to maintain surveillance over the entire North Atlantic trade routes, so that the notorious Atlantic Gap, within which the U-boats could congregate out of range of Allied aircraft, disappeared. By the spring of 1943, the Allies had also created five special support naval groups that included escort carriers as well as destroyers. These operated independently of the convoys and could be directed by aircraft to the exact position of the U-boats and destroy them. Finally, with the output of the US naval shipyards becoming available, the convoy escorts were also strengthened.

JUST AS HENRY FORD'S mass-production plants of the second and third decades of the twentieth century were an engineering bridge between the earlier achievements of the American system of manufacture and the later elaboration of the mass-production model during the remainder of the twentieth century, so the US Air Force's SAGE air-defense system developed in the 1950s linked the management control systems of the Second World War to the corporate control systems that began to emerge in the 1960s and with which in a mature form we are living today.[8] Standing for Semi-Automatic Ground Environment, SAGE as a "total information system" combined the geographical reach of the systems deployed in the Battle of the Atlantic with the speed of command and control of those of the Battle of Britain in the summer of 1940. The big difference was that SAGE was a heavily automated system relying on the most advanced computers of the time.

Although formally SAGE belonged to the military-industrial complex that President Eisenhower warned about in his farewell address of January 1961, it had its origins in a group of scientist-managers, based mostly at MIT, who nurtured the project and brought it to fruition. Its equivalent of J. Robert Oppenheimer was Jay Forrester, director of the Lincoln Laboratory at MIT in the early 1950s and a creative force for SAGE both scientifically and bureaucratically with the program's dealings with the Pentagon, the air force, and civilian IT companies participating in SAGE, such as IBM. From the late 1940s through the 1970s, this military-academic complex would dominate the research, development, and deployment of large-scale computerized management systems. Its influence on the private corporate systems that began to emerge in the 1960s and 1970s was overwhelming. It is literally impossible to make sense of today's CBSs without allowing for this history.

As a case history in the economics of innovation, the SAGE project challenges the claim of market neoliberal economics that the private sector is invariably the font of significant technological change. The technological breakthroughs achieved by SAGE during its decade-long gestation from the late 1940s to the late 1950s were prodigious, but the indispensable catalysts were Forrester's group of computer scientists at MIT and the generous funding provided by the air force, with private corporations such as IBM, Raytheon, and Bell Telephone playing a subordinate role as contractors for particular segments of the system. As the main driver of innovation, a sense of urgency about the need to defend the United States from Soviet nuclear attack substituted for the pressures of market competition.

When from the 1960s onward the private sector set about developing its own large-scale management systems, relying on its own expertise and resources, progress was by comparison slow.

As a dual air-defense and command-control system, SAGE performed four functions: the gathering of raw information about the attacking enemy, the filtering of this information by computers to come up with a clear vision of the developing attack, the deployment of the defending forces to engage the attacker, and the guiding of these forces to achieve final contact with the enemy. The R & D phase of SAGE, undertaken mostly at Forrester's Lincoln Laboratory at MIT, achieved three critical breakthroughs that made possible the information and the command-and-control components of SAGE, with its elaborate physical structure deployed over thousands of square miles of the United States and Canada.

Under the overall control of the Joint Chiefs of Staff's Continental Air Defense Command, SAGE's own command structure divided the continental United States and Canada into eight sectors, each with a combat center, and below them thirty-two subsectors, each with a direction center. At the direction centers filter rooms with computers made sense of the mass of information about the developing Soviet attack coming in mostly from radar stations. This information was then sent to the combat centers at sector level, where it was put together to form an overall view of the military situation within the sector so that decisions could be made about the deployment of the defending US forces. Once the attacking forces had been identified and the defending force assigned—something that required human intervention—the calculation of the flight path of the defending forces to the target was automated.[9]

It was in the detailed performance of these tasks that the breakthroughs achieved by Forrester's team at MIT came into their own and revolutionized both the nature of air defense and the role of computers in large-scale management systems. The first of these breakthroughs was the ability of the system's computers at the direction centers to gather the electronic information provided by radar and to translate it at high speed into a digital form that could be communicated to other computers; the second was this ability of computers relying on telephone lines to communicate with one another, however distant; and the third was the ability of the system's computers to calculate automatically the flight path to the target of the defending aircraft and missiles.

Like Ronald Reagan's Strategic Defense Initiative of the early 1980s, SAGE in any realistic estimate would never have been able to provide an impenetrable dome of defense for the continental United States. Experts at the time predicted that between 30 percent and 40 percent of the attacking Soviet bombers would get through. Also SAGE became obsolete even before it became operational because the Soviet Union deployed its first ICBMs in 1958, against which SAGE could provide no defense. Yet SAGE occupies a pivotal position in a seventy-year history linking the predigital information systems of World War II to today's most advanced and ambitious CBSs. SAGE anticipated by forty years the automation of business process as promoted by Hammer and Champey and the reengineers, and also the construction on this reengineered base of much more ambitious management information systems as the civilian equivalent of SAGE's command and control.

With the ability of computers to communicate with one another irrespective of distance, the size of an organization and the distance of its constituent parts from one another ceased to be limiting factors in the creation of complex, automated management systems—whether military or corporate. Again, with the ability of computers to gather raw information and to translate it at high speed into forms needed by senior managers—whether air force generals or corporate CEOs—these human actors acquired panoptic information about the condition of corporate or military processes in real time and, with it, the power exercised by the sector commanders in SAGE's direction centers to respond immediately to adverse developments, also in real time.

Beyond the technological breakthroughs achieved with SAGE that shaped the systems both military and civilian that came after it, there was another dimension to this influence that SAGE exerted both as a system designed for the military and as one designed for the battlefield and for the *nuclear* battlefield in particular. The SAGE technologies were embedded in systems designed to cope with the extreme, apocalyptic circumstances of nuclear war, and this singular role imposed extreme biases on the system that in turn showed up in the corporate control systems spawned by SAGE and its military successors. The existence of this transfer is one of the reasons the early history of civilian CBSs makes no sense without making allowances for this military dimension. The scope of the transfer also points to a strong affinity between military and corporate cultures.

With SAGE and its successors, there was an extreme bias toward *total* information about the developing nuclear battlefield, because a failure to find and destroy even a handful of attacking bombers would have catastrophic consequences for the US civilian population; there was a bias toward giantism, given the system's need for continental reach; there was a bias toward speed, because the central military process of searching, finding, and destroying the enemy had to take place under the extreme time pressure exerted by the inexorable movement of Soviet bombers toward American cities; and there was a bias in favor of hierarchy and centralized command, because the decisions of senior officers about the conduct of the battle had to be implemented immediately and exactly as ordered. There was even a bias toward the deskilling of the system's "front-line workers"—the pilots flying the intercepting fighters to their targets—because in reaching their targets, the pilots were subordinate to automated control systems on the ground.

In this history of relations between the military and civilian versions of CBSs, there is a moment of technology transfer between the two that must rank among the seminal events of contemporary American economic and business history. This was the moment in 1962 when the technologies of SAGE, the Semi-Automatic Ground Environment, crossed over into the civilian world and became SABRE, the Semi-Automatic Business Research Environment. SABRE was the automated airlines reservation system developed at huge cost by IBM, the veteran of SAGE, for American Airlines and allowed agents at specially designed consoles "to interrogate a central computer in order to review flight availability and make reservations."[10] SABRE became the first civilian Computer Business System in the form that we know them today.

As the raw material of their respective systems, the automated search for seats on a flight from New York to Chicago does not quite rank with a search for the flight path of a Soviet bomber en route to Chicago. But with the gathering and filtering of information from tens of thousands of separate information sources, dispersed over thousands of miles from one another throughout North America, SABRE reproduced the giantism of SAGE in a civilian setting. It also followed SAGE in becoming a management information system, eventually providing senior American Airlines managers with real-time information about patterns in the demands for flights according to cost, time, type of aircraft, and destination. It took another thirty years for these CBS technologies to become a dominant force in American business. But the seeds of this dominance were planted at a time when John F. Kennedy was president and Nikita Khruschev was smuggling Soviet missiles into Cuba.

8

THE NUCLEAR HALF

THE BATTLE OF THE ATLANTIC ENDED IN VICTORY FOR THE ALLIES, as did two other system wars of World War II, the Battle of Britain in 1940 and the air war over Germany from 1941 onward. But two other military case histories that are a part of this series, the Vietnam War and the management of the US strategic nuclear forces, especially in the decade 1975–1985, exemplified error on a scale commensurate with the scale of management giantism itself. They are included in this history not because the theorists and reengineers of CBSs have necessarily learned the lessons of these errors and guarded against them. There is no evidence that they have. The two problematic histories are included because the civilian CBS project itself, conceived as a ninth case history of management giantism, contains strong elements of folly and error, threatening our economic and social well-being. It might therefore be wise to review at least one of these dysfunctional histories so as fortify ourselves against allowing the civilian designers of CBSs to commit similar errors now.

The architecture of what I will call the decade of nuclear dysfunction (DND) between 1975 and 1985 rested upon an extremely elaborate and detailed statistical analysis of the nuclear capabilities of the United States and the Soviet Union and equally detailed calculations about what would happen if these weapons were used in anger. These simulations of nuclear war originated in the laboratories of the Department of Defense and of such semiofficial bodies as the RAND Corporation. But in the context of the DND, the most significant users of the data were not the uniformed military or civilian officials in Washington, but the theorists of nuclear alarm, whose ideas by the late 1970s and early 1980s encompassed much of the civilian and political leadership of the United States.

With the theorists of the DND, the role of this accumulated statistical database was to anticipate the possible course of events in a nuclear war between the United States and the Soviet Union. There had to be an element of abstraction in this because the anticipated events had never occurred, except at Hiroshima and Nagasaki, where the "modest" scale of operations could not be a model for what would happen in a full-scale nuclear exchange. However, the war-gaming scenarios created by the theorists of the DND were still perversely abstract because they took place in a disembodied realm where precise statistical calculations of marginal advantages allegedly possessed by one side or the other—in "throw weight," number of warheads, accuracy of warheads, or sheer destructive megatonnage—supposedly yielded significant and tangible advantages that would have a practical impact on how the stronger party would conduct its relations with the rest of the world.

The nuclear war gaming of the period was frightening in its detachment from a reality in which the United States—supposedly the weaker, inferior side—alone possessed more than *ten thousand* nuclear weapons that could reduce the entire Soviet Union and indeed the whole world to dust hundreds of times over, and with the Soviet Union being able to do the same to the United States. The theorists used the cool language of game theory to describe their limited, calculated nuclear exchanges in which tens of millions of American or Soviet citizens would certainly perish—20 million became the preferred estimate for the number of lives that the Soviet leadership would be prepared to sacrifice in launching its first strike. This trivializing of nuclear war is a backdrop to the whole body of nuclear war gaming that took over in Washington by the late 1970s and early 1980s. It is this pathology of abstraction that links the nuclear theorists of the DND forty years ago to the operators of corporate CBSs today.

One of President Dwight D. Eisenhower's great legacies to the Republic was the concept of "nuclear sufficiency," the idea that once the nuclear arsenals of the superpowers had reached a certain level, marginal shifts in the strength of either side were of no military significance and could "only make the rubble jump," to use the language of the Cold War era.[1] It was Eisenhower's unique qualifications as former supreme commander and war hero that gave him the authority to make nuclear sufficiency the foundation of strategic policy and to overcome the endemic resistance of the US military, especially the air force and its allies in Congress, to any policy that might curb their appetite for new and more costly weapons systems.

Despite John F. Kennedy's warnings during the presidential campaign of 1960 of a "missile gap" favoring the Soviet side, abandoned once he was in office and saw the classified documents, nuclear sufficiency remained a foundation of US strategic thinking during the Kennedy, Johnson, Nixon, and Ford administrations. As long as it was believed on both the American and the Soviet sides that the strategic forces surviving a first strike by the other were strong enough to deliver a devastating retaliatory strike against the adversary, then the balance of forces between the two superpowers was thought to be stable—the doctrine of mutually assured destruction (MAD). In a 1982 interview with journalist Robert Scheer, a chastened, post-Vietnam Robert McNamara evoked the realities of nuclear war to explain the logic of MAD:

> To try and destroy the 1,054 Minutemen [US land-based missiles], the Soviets would have to plan the ground burst of two nuclear warheads of one megaton each on each site. That is 2,000 megatons, roughly 160,000 times the megatonnage of the Hiroshima bomb. What condition do you think our country would be in when 2,000 one megaton bombs ground burst? The idea that, in such a situation, we would sit here and say, "Well, we don't want to launch against them because they might come back and hurt us," is inconceivable. . . . [It is] too incredible to warrant serious debate.[2]

With strategic stability underwritten by MAD, it was possible to negotiate limitations on the growth of each side's strategic nuclear arsenals, something achieved by the Nixon administration in May 1972 with the signing of the Strategic Arms Limitation Treaty (SALT) in Moscow and by the Carter administration with the signing of the SALT II Treaty in June 1979, also in Moscow—although the treaty was never ratified by the US Senate. However, strategic stability and the doctrines that underpinned it began to erode from 1975 onward, a casualty of Watergate and Nixon's decline and fall and of Henry Kissinger's loss of influence as secretary of state during the Ford presidency.

The erosion would accelerate during the Carter administration (1977–1981) and come to a head during Ronald Reagan's first term (1981–1985), among the most dangerous periods of the entire Cold War. From the mid-1970s onward, the initiative on strategic nuclear issues lay increasingly with a community of hard-line theorists on nuclear war whose most influential members were Paul H. Nitze, the veteran State Department official, and Richard Perle, an aide to Senator Henry Jackson (D-WA) with an encyclopedic knowledge of strategic nuclear matters. Along with Perle and Nitze, the group included Eugene Rostow, an undersecretary of state in the Johnson administration and (inappropriately) director of the Arms Control and Disarmament Agency under Ronald Reagan; Richard Allen, a National Security Council official from the Nixon administration who became Reagan's first national security adviser; Richard Pipes, a specialist in Russian and Soviet history at Harvard and Soviet expert at the NSC under Reagan; Fred Iklé, director of ACDA in the Nixon administration and undersecretary of defense for policy under Reagan; and, from about 1976 onward, when he ran against Ford for the Republican nomination, Ronald Reagan himself.

Perle, Nitze, and their group saw the SALT Treaty negotiated by Nixon and Kissinger as fatally flawed, because it contained loopholes that allowed the USSR to achieve what they claimed was nuclear superiority over the United States in the form of a disarming first strike against the United States' land-based ICBMs. Neither Perle nor Nitze ever actually stated that the Soviet side, emboldened by the possession of nuclear superiority, would out of the blue launch a first strike against the United States. But both certainly believed, and frequently stated, that because the Soviet Union would in fact emerge the victor in a nuclear exchange with the United States, an awareness of this Soviet superiority would decisively influence the conduct of both superpowers in the context of a major confrontation between them. In Richard Perle's words: "I worry about an American President feeling he cannot afford to take action in a crisis because Soviet nuclear forces are such that, if escalation took place, they are better poised than we are to move up the escalation ladder."[3]

By the beginning of Carter's presidency in 1977, the nuclear ultras had come to dominate the nuclear thinking of the Republican Party and marginalized any lingering attachment to "nuclear sufficiency" as espoused by Richard Nixon and Henry Kissinger. Looking back at this period with nearly forty years of hindsight, it is a puzzle and a scandal that the Perle-Nitze theories carried all before them in the mid- and late 1970s and shaped the strategic policies of the Carter and Reagan administrations. How could this have happened—how could theories that Nobel Prize–winning physicist Hans Bethe described as "crazy" have become by the end of the 1970s the received wisdom of much of the US civilian and military leadership and a dominant influence on their policy making?[4]

It was a spectacular example of how policy making can become divorced from its historical context and how data gathered and analyzed by information systems can become mixed in with judgments that are very far from scientific, with the whole mixture taking on a spurious scientific authority once jumbled up by clever and often unscrupulous bureaucratic and public policy operators. But the theorists of nuclear war were very successful during their prime decade, and this poses an awkward question: how and why did the guardians of nuclear sanity allow this to happen? In his book *The Closed World: Computers and the Politics of Discourse in Cold War America,* Paul Edwards of the University of Michigan defines the closed world as a system of thought that, though seemingly systematic and coherent, excludes important segments of the reality it seeks to explain.[5] The war-gaming theories of the nuclear ultras were archetypal examples of closed-world thinking and certainly among the most dangerous of recent times. The reality they excluded, and by virtue of which their thinking was closed, was the unimaginable horror of what would actually happen to the world if the nuclear exchanges described by the ultras actually took place and if even a tiny fraction of the thousands of nuclear weapons in the hands of the superpowers were used in anger.

There were four main components to this closed-world thinking on nuclear war that became dominant in the United States during the DND: first, a reliance on advanced information systems to simulate the consequences of nuclear exchanges for each side's strategic forces, populations, cities, and economic and military targets, with precise quantification of these effects; second, a reliance on game theory to predict, again with confidence, exactly how each side's leadership would behave as the nuclear exchanges between them had gotten under way; third, the evoking of a continuing fear of the Soviet Union and mistrust of its leaders to lend plausibility to the claim that the Soviet leadership would behave with suicidal recklessness in launching or threatening to launch a first strike against the United States; and fourth, the avoidance of any detailed discussion of the real-world effects of large-scale nuclear exchanges involving hundreds if not thousands of nuclear weapons—the most "closed-world" aspect of this nuclear theorizing.

Early in 1977 I encountered this closed-world thinking in the flesh and from one of the best possible sources. I was a reporter in New York at the time and had begun writing about the role of these nuclear questions in Soviet-American détente. I called up Paul Nitze at his Washington office. Nitze answered the telephone immediately, as if he had been just sitting there waiting for reporters to call up and ask about the Soviet threat, which he proceeded to describe in the bleakest terms. The great Clint Eastwood himself could not have done a better impersonation of a veteran Cold Warrior raging against the menace from the Kremlin. I was struck by the brisk confidence with which Nitze quantified the outcome of nuclear exchanges: the percentage of US ICBMs that would survive, the damage the US retaliatory strike could inflict on Soviet "hard" targets. I was also struck by Nitze's confidence in anticipating exactly how the leadership of the two sides, and especially of the United States, would behave as Armageddon unfolded.

Nitze stated again, with considerable assurance, that the US leadership would be paralyzed by the threat of a Soviet first strike because, once the strike had taken place, the United States would lack the accurate counterforce weapons to respond in kind against Soviet hard targets. It would not dare escalate the level of violence and target Soviet cities out of fear that the Soviet leadership would then target American cities. So paralysis would ensue, or, in Ronald Reagan's version of nuclear paralysis, "[The Soviets] don't want . . . a direct confrontation with us . . . until they have such an edge that they could realize their dream of perhaps taking us by telephone. Then we would have no choice left except surrender or die."[6] Absent from both Nitze's and Reagan's utterances was any acknowledgment that tens of millions of American and Russian lives would be at high risk as these nuclear scenarios unfolded.

From about 1977 onward, I watched with mounting dismay as Perle, Nitze, their allies in Congress, and their propaganda vehicle, the Committee on the Present Danger, steadily gained ground and the Carter administration, like the US leadership in Nitze's nuclear war-gaming scenarios, seemed paralyzed and unable or unwilling to put up a fight, even though the administration was committed to arms-control negotiations with the Soviet Union and actually signed the SALT II Treaty in June 1979. The fatal error of the Carter administration during those years was its failure to provide a "real-world" account of nuclear war, nuclear sufficiency, and the sufficiency of the US nuclear triad in particular. Instead, it allowed itself to be dragged down into the murk of closed-world theorizing where it was no match for the propaganda and pseudoscientism of the ultras.

These latter managed to combine the four elements of closed-world thinking in ways that combined cool, scientific rationality with crude demagoguery. The pseudoscientism comprised the precise quantification of the effects of nuclear war supposedly yielded by information systems and the equally precise anticipation of the conduct of the opposing Soviet and American leaders yielded by game theory. The demagoguery linked the crude brutalism of the Soviet heavy missiles with the crude brutalism of the Soviet leaders themselves—desperadoes who would stop at nothing in their quest for world domination—and with the heavy missiles as a kind of physical embodiment of the Soviet leaders.

The Carter administration lacked anybody with the authority and intellectual stamina to make the case for nuclear sufficiency, as the pre-Watergate Nixon and Kissinger perhaps could have done. Two of its most senior officials, Secretary of State Cyrus Vance and Secretary of Defense Harold Brown, were high-echelon technocrats with modest political and rhetorical skills. Its third lead official, Zbigniew Brzezinski, the national security adviser, was the closest of the three to the ultras, though for reasons that had to do as much with his anxieties about the Soviet presence in remote corners of Africa as about the threat of Soviet ICBMs. Presiding over them all was Jimmy Carter who, despite his background as a nuclear engineer in the navy, lacked the confidence and experience to take on the doomsayers and dithered between the conflicting advice he was getting from within his own administration.

The proof of this was the MX affair, a now-forgotten but at the time bizarre offshoot of the nuclear war gaming in Washington. As conceived by the Carter administration, the MX missile was a giant ICBM that would not be based in fixed missile silos and so would not be vulnerable to a Soviet first strike. The MX would instead roam the western states in specially constructed roads or railroads, either overground or underground, so that their exact position would not be known to the Soviet side and they could not be targeted in a first strike. But how could this concealment be reconciled with the terms of the SALT II Treaty, which imposed precise limits on the number of permitted US ICBMs and to which the Carter administration was committed?

How also could the Russians be sure that the United States was not violating the treaty by hiding some rogue MXs in the dark recesses of the networks, especially if they were buried underground? The Carter administration then came up with the idea of holding the Cold War equivalent of the Day of Judgment for the MX missiles, when they would be counted. At a time agreed with the Soviet side, each MX missile would be positioned at one of its firing points on their networks so that the flaps concealing the firing points could be lifted and the Soviet spy satellites could peer down from their orbits and verify that there were indeed no rogue MXs hiding in the tunnels.

Because the state of Utah was one of the chosen sites for the MX network, the elders of the Church of Jesus Christ of Latter-day Saints, the Mormons, did the Republic a singular service when they declared in 1981 that they did not want these weapons of mass

death contaminating their holy land. The search for a western mobile home for the MX missile was finally called off when Senator
Paul Laxalt of Nevada, a close ally of Ronald Reagan, declared also
in 1981 that his state did not want to play host to the MX either.[7]
But by the time Senator Laxalt vetoed the MX networks in Nevada,
Ronald Reagan was president, and so someone who from the mid-
1970s had been a bellicose proponent of the impending strategic
doomsday wielded executive power in Washington. Indeed, by then
Paul Nitze and Richard Perle themselves were both senior officials
of the Reagan administration.

The failure of the Carter and Reagan administrations to find
a safe home for the MX missile was not the end of this comedy of
nuclear error, because it was followed by another and even more
pivotal development. This was Ronald Reagan's announcement, in
his radio address of March 23, 1983, that he was committing the
United States to the search for a total space-based defense against
Soviet ICBMs—the Strategic Defense Initiative, or "Star Wars." In
his opening pitch for SDI, Reagan used the language of idealism,
utopia, and American technological exceptionalism: "Wouldn't it
be better to save lives than to avenge them? Are we not capable of
demonstrating our peaceful intentions by applying all our abilities
and our ingenuity to achieving a truly lasting stability? I think we
are—indeed we must!"[8]

Reagan invited his audience to look forward to a nuclear-free
world: "What if free people could live secure in the knowledge that
their security did not rest upon the threat of instant U.S. retalia-
tion to deter a Soviet attack?" Reagan conceded that this was a "formidable technical task, one that may not be accomplished before
the end of this century." Yet it was a goal worth pursuing, and here
Reagan could draw on American technological exceptionalism and

memories of the Manhattan and Apollo Projects: "Current technology has attained a level of sophistication where it is reasonable for us to begin this effort. It will take years, probably decades, of effort on many fronts. There will be failures and setbacks, just as there will be successes and breakthroughs."

Reagan's reliance on the language of idealism to sell SDI to the American people was a stroke of genius because it had the effect of detaching SDI from the geopolitical and strategic contexts in which it was conceived, which were fraught and dangerous, and locating it instead within a pure and benign moral sphere actually beneficial to the American people. Lou Cannon, Reagan's two-time biographer and someone who has followed him and has written about him more than anyone else alive, accepts this view of Reagan as visionary and romantic: "Reagan totally believed in the science-fiction solution he had proposed. . . . Reagan was convinced that American ingenuity could find a way to protect the American people from the nightmare of Armageddon. As he saw it the Strategic Defense Initiative was a dream come true."[9]

This view of Reagan ignores the deep entanglement of the Reagan administration and of the president himself in "closed-world" thinking about nuclear war, where the president and his closest advisers believed that "nuclear superiority" was a meaningful concept and that the Soviet Union possessed it. The administration's determination to deny the USSR this advantage and reassert American primacy took the form of an official commitment to acquire the means to fight and win a nuclear war. This was spelled out in the administration's national security decision document of May 13, 1982:

> Should deterrence fail and strategic nuclear war with the USSR occur the United States must prevail and be able to force the

Soviet Union to seek earliest termination of hostilities on terms favourable to the United States. . . . [T]he United States must have plans that assure US strategic nuclear forces can render ineffective the total Soviet military and political power structure . . . and forces that will maintain, throughout a protracted conflict period and afterward, the capability to inflict very high levels of danger against the industrial/economic base of the Soviet Union.[10]

Here was the ultras' closed-world thinking about nuclear war enshrined in official US doctrine. In the shadow of this resolve to fight and win a nuclear war was Paul Nitze's confidence that nuclear war could be managed, its damage quantified and limited, and the decision making of its participants confidently anticipated. In this closed world where the ability of each side to destroy the strategic forces of the other was the measure of victory or defeat, the possession by each side of highly accurate "counterforce" weapons was the measure of military preparedness. The nuclear modernization of the Reagan administration, well under way by the time of the Star Wars speech in March 1983, was heavily concentrated on the acquisition of these counterforce weapons.

The MX missile, assuming a home could be found for it, was designed as a silo-busting weapon. The Trident SLBM gave the United States the capacity to hit Soviet hard targets from the oceans. The cruise missile, launched from US bombers or submarines, was also a highly accurate counterforce weapon against hard targets, whether missile silos, communication networks, or military bases. Reagan concealed the role of SDI in this war-fighting strategy with the language of utopia, but SDI was nonetheless a war-fighting weapon. To see why requires the heroic assumption that SDI could be made to work—but an assumption no less heroic than the assumption that the destruction of nuclear war could be controlled and measured or that the decision making of the antagonists in the heat of battle confidently predicted. In the surrealism of the closed world, figments of science fiction could fast mutate into hard objects already up there roaming in space. In such a role SDI was as much a Soviet missile killer as the MX and the Trident missile and could make its own special contribution to nuclear victory.

THE RISE OF Mikhail Gorbachev as general secretary of the Soviet Communist Party in 1985 was the development that more than any other brought the DND to an end and launched the sequence of events that in 1989 culminated in the coming down of the Berlin Wall and the end of the Cold War. But where does that leave the theories and practices of the American side of the DND, embraced by the Reagan administration from 1981 onward? For true believers such as Richard Perle, the winding down of the Cold War in the second half of the 1980s was a vindication of the aggressive strategy he and his cotheorists been advocating since the 1970s, with the threat to escalate the nuclear arms race into space with SDI finally convincing Gorbachev that the Soviet Union could compete with the United States only at a cost of economic collapse, and so forcing him to throw in the towel.

But the historical record does not bear this out. In one of the great ironies of Soviet history, it was Andrei Sakharov, the world-renowned physicist, inventor of the Soviet H-bomb, Nobel Prize winner, and the Soviet regime's most formidable critic during the Brezhnev era, who, after his return to Moscow from exile in Gorki in 1986, took the lead in persuading Gorbachev that SDI was a strategic white elephant and that Gorbachev should not make the negotiations on the reduction of strategic weapons conditional on the United States' giving it up. At a conference on disarmament held in Moscow in mid-February 1987 and attended by Gorbachev, Sakharov argued that SDI would never be militarily effective against an opponent with a nuclear arsenal the size of the USSR's; rather, it would be a "Maginot line in space."[11]

Even before his encounter with Sakharov, Gorbachev had, like Eisenhower before him, come to believe that the vast nuclear arsenals on both sides guaranteed nuclear sufficiency; that the marginal advantages in throw weight, megatonnage, and warhead numbers fixated upon by the American nuclear theorists were meaningless; and that a basis for negotiations between the two sides already existed. It was this logic that led Gorbachev to propose radical reductions in the two sides' strategic arsenals—even their abolition—before and during the Reykjavík summit with Reagan in October 1986.[12]

The coming of Gorbachev was providential, and to see why one has only to imagine what would have happened if Gorbachev's orthodox Soviet predecessor, Yuri Andropov, had lived as long as his predecessor, Leonid Brezhnev, and so remained in office for at least the years of Ronald Reagan's second term (1985–1989). Reagan would then have had to follow Nixon and Kissinger in elaborating a complex diplomacy to deal with the endemic ambiguities of Soviet conduct. As it was, Reagan's Star Wars speech left US strategic doctrine and the United States' nuclear diplomacy with the Soviet Union in a state of extreme confusion. The definitive account of this post-SDI chaos in Washington is to be found in Frances FitzGerald's seminal work, *Way Out There in the Blue: Reagan, Star Wars, and the End of the Cold War.*[13]

The Star Wars speech unleashed a classic Washington deadlock with on one side Secretary of Defense Caspar Weinberger and Richard Perle, the most influential theorists of the DND in the Reagan administration and pushing for the development and deployment of SDI as a weapon of American nuclear superiority. Opposing them were Secretary of State George Shultz and a newly moderate Paul Nitze, not wanting to close off the possibility of arms-control nego-tiations with the Soviet side and willing to use SDI as a bargaining chip to that end. As secretary of state, Shultz was also exposed to the concerns of the United States' European allies, who would be un-protected by SDI and feared that its development and deployment would prolong the Cold War.

What was Ronald Reagan's role in all this? The portrait that emerges from FitzGerald's exhaustive account is of a president who lacked the intellectual stamina and the taste for personal confron-tation needed to resolve this Washington deadlock one way or an-other. For at least the two years following the Star Wars speech, US nuclear strategy and nuclear diplomacy were in a state of drift, and it was only with Gorbachev's consolidation of power in the Kremlin from about mid-1985 onward that this statis ended. But in respond-ing to Gorbachev, was Reagan acting in the spirit of Weinberger and Perle, acknowledging what he believed to be the surrender of the Soviet adversary, or had Reagan in old age embraced the wisdom of Dwight Eisenhower and accepted the logic of nuclear sufficiency?

Reagan's official biographer, Edmund Morris, became so con-fused about the identity of the real Reagan that in his book *Dutch: A Memoir of Ronald Reagan,* he resorted to the desperate measure of inventing fictional encounters between himself and Reagan in the hope that the true Reagan might somehow emerge.[14] It is perhaps wiser to concede that the mystery of Reagan's motives will never be unraveled and that the mystery has gone with him to his grave. What is certain is that without Gorbachev's lead, Reagan would have been hard-pressed to put forward a coherent vision of how to end the Cold War, let alone impose this vision on his fractious administration. But if Gorbachev was the chief architect of the Cold War's end, Ronald Reagan was, whatever his motives, Gorbachev's partner, and for that we are in his debt.

9

THE CHINESE MODEL

IN NOVEMBER 2012 THE ORGANIZATION FOR ECONOMIC COOPERAtion and Development in Paris (OECD) forecast that before the end of 2016, China would overtake the United States as the world's largest economy.[1] This would perhaps be the most significant reshaping of the global balance of power since the eclipse of Europe and the rise of the American and Soviet superpowers in the aftermath of World War II. The Chinese economy is already endowed with many of the sophisticated attributes of a first-world capitalist economy. China plays a leading role in global financial and currency markets with its heavy purchase of US government debt and its manipulation of its currency, the renminbi (RMB), for its own trading advantage. It is a source of inward investment for developed and developing economies of the third world, especially in Africa. The growth and slowing down of its gross domestic product are leading determinants of global economic health and a major news story in the United States and Europe.

This rise of a capitalist China is entwined with a paramount question of political economy that receives much less attention in the West: how does the continuing hegemony of the Chinese Communist Party, undiminished by the post-Mao economic reforms, shape the nature of Chinese capitalism itself? It is here that the thesis of this book can provide insights into why the CCP has acted as it has: why the party believes that it can achieve the best of both worlds with a system of capitalist production that succeeds in the global marketplace *and* can subject Chinese labor to a regime of command from above that can achieve the political neutering of this fast-growing proletariat, and no less effectively than the mechanisms of bureaucratic control of the Soviet-era command economy that ruled in China from 1949 until the coming of Mao's Cultural Revolution in 1966.

In this age of outsourcing, it might be said that the CCP has outsourced the tasks of workplace control to the predominately East Asian enterprises—Taiwanese and Hong Kong Chinese, Japanese, and Korean—that have driven the explosive growth of Chinese manufacturing in its two dominant industrial hubs, the Pearl River Delta (PRD), centered on Guangzhou and Hong Kong itself, and the Yangtze River Delta, centered on Shanghai. The tens of thousands of factories that these companies have set up in the two zones, employing tens of millions of Chinese workers, are overwhelmingly mass-production plants turning out textiles, clothing, footwear, consumer electronics, desktop and laptop computers, mobile phones, and automobiles.

This is a world where the Ford-Taylor model of industrial production, as mediated by Japanese industrial practice, reigns supreme. These plants are also operating in a political and social context that subjects their labor force to a degree of workplace pressure and control that goes beyond anything to be found in the United States or Europe, though the regimes now in place at Walmart and Amazon get uncomfortably close. Meanwhile, higher up the corporate food chain, US consulting firms are earning very substantial fees as they graft Computer Business Systems and Corporate Panoptics onto the management structures of Chinese state enterprises, including those recently privatized. But these systems, with their military origins, embody, as we have seen, a corporate variant of centralized control that mimics the party's own. The CCP has outsourced with care.

Here then is the transfer to Chinese soil lock, stock, and barrel of the dual systems of management control that define contemporary American capitalism: the transfer of Business Process Reengineering, with its shaping of horizontal business processes in both manufacturing and services, and Corporate Panoptics, with its empowerment of top management with the electronic representation of the corporate organism in its entirety and in real time. I have argued that the combination of these two control regimes—one pushing up from the corporate bottom, the other pushing down from the corporate summit—chiefly benefits top management. But does this empowerment of a managerial elite also create in a Chinese setting a competing center of power that will eventually bring the hegemony of the CCP to an end with a Chinese version of the collapse of Soviet power in the early 1990s?

There are strong reasons to believe that this will not happen and that there exists a de facto alliance between the CCP and the managerial elite that is as close as "lips and teeth," as Mao once said of the alliance between China and North Korea at the time of the Korean War. But what managerial elite? The managerial vanguard that has driven China's rise as an economic great power is, paradoxically, both offshore *and Chinese.* Although the offshore economies of greater China, Taiwan, and Hong Kong are small by global standards, they have driven Chinese industrialization on a scale and with a ferocity not seen since the Industrial Revolution, unleashing a mass migration of labor from the Chinese villages to the cities on a scale evoking folk memories of the ancient world—the building of the pyramids and of the Great Wall itself. This management compact with the CCP also provides the framework within which other foreign investors in China—Japanese, Korean, European, and American—can conduct their Chinese business.

The benefits the CCP has bestowed on their offshore business compatriots and these foreign investors are very considerable: access to virtually unlimited supplies of cheap Chinese labor, access to the huge Chinese domestic market, a legal code that underwrites managerial hegemony in the workplace and effectively bars the formation of independent labor unions, the licensing of provincial and city party organizations to subsidize inward investors with tax concessions and land grants, and freedom from state and party interference in the detailed planning and organization of production.

In return, the business elite, with greater China to the fore, has provided the wherewithal for the rise of China as an economic and diplomatic great power and has been a surrogate for the CCP in the taming of China's new industrial proletariat. On the business side of this grand bargain, why should such a favored managerial oligarchy bite the hand that has so generously fed it, and why would it risk unleashing forces in China that might not be able to keep order and deliver favors with the assurance of the CCP? This political and economic logic applies equally well to the management of indigenous Chinese companies becoming part of China's global economy, such as the telecommunications equipment maker Huawei, or the computer makers Lenovo and BYD Electronic of Shenzen—the latter in

March 2013 beating out the leading Taiwanese manufacturers Hon Hai and Quanta to be the maker of Hewlett-Packard's Slate 7 tablet.[2]

Although Hong Kong and Taiwanese businesses have driven Chinese industrialization, they have in their own industrialization drawn heavily on the Japanese variant of the Ford-Taylor model, introducing these methods in China as they have set up factories there from the 1980s onward. As a financial journalist for the Hong Kong–based *Far Eastern Economic Review* and the London *Financial Times* in the late 1960s and early 1970s, I had watched the early stirrings of this new Japanese coprosperity sphere in East and Southeast Asia. Japanese corporations such as Panasonic and Sony began building plants in Hong Kong, Taiwan, and South Korea and then in the more developed Southeast Asian economies such as Malaysia and Thailand. There they turned out basic electronic goods such as radios, tape recorders, and televisions.

In postwar Japan early generations of industrial workers were drawn mostly from the rural areas, and Japanese industrialists such as Eiji Toyoda had to adapt their methods to allow for the limited skills of a peasant workforce in much the same way that Henry Ford had done with immigrant laborers in Detroit. This was the setting for Toyota's reliance on Scientific Management for the development of its production system, now widely diffused in the US auto industry.[3] These postwar lessons were also applied to Japanese companies active throughout East and Southeast Asia. Work on the Asian assembly lines was reduced to its simplest elements. Time and motion studies were rigorously carried out, and teams of supervisors closely observed the line. Workers were allowed to suggest ways in which the line might be speeded up, but management always retained full powers to decide whether and how these suggestions might be used.

Although always adhering to Japanese industrial culture, the later industrialization of Hong Kong, Taiwan, and also Korea has followed differing paths, and this has had a major impact on China's own industrialization. The Taiwanese and Korean governments have helped develop an R & D infrastructure in IT that over time has enabled their leading IT companies to equal and then overtake their Japanese competitors, notably Samsung in consumer electronics, the Taiwan Semiconductor Manufacturing Company, and the Taiwanese United Microelectronics Corporation in customized microprocessors. In Hong Kong this industrial upgrading never took place. The Hong Kong government did not support the development of an R & D base comparable to Korea's or Taiwan's and capable of sustaining major product innovations. The result has been that Hong Kong industry has gone on turning out lower-tech, labor-intensive consumer electronics and electric goods. Yet while Hong Kong may have been a minor industrial power, it is a world-class financial power, and with the opening up of South China to foreign investment, local Hong Kong industry and global Hong Kong finance came together with spectacular results.

By the end of the 1990s, there were forty thousand Hong Kong companies in the Pearl River Delta adjacent to Hong Kong, employing 10 million workers, more than Hong Kong's entire population. Mirroring Hong Kong's industrial profile, the zone became the world's leading exporter of lower-tech consumer electronics and electrical goods. By the millennium the Pearl River Delta produced 79 percent of China's telephones, 43 percent of its video recorders, 35 percent of its VCR players, 88 percent of its electrical fans, and 35 percent of its color televisions. When expanded to include Macao and Hong Kong itself, the PRD became the world's sixteenth-largest economy and its tenth-largest exporter. Hong Kong investment helped transform Shenzen, just over the border from Hong Kong's

New Territories, from a provincial city of 300,000 to a metropolis of nearly 8 million.[4]

Taiwanese investment in China is more recent and more evenly divided between South China—the PRD—and central, coastal China, the Yangtze River Delta, with Shanghai as its hub. It is also focused on more sophisticated IT products, with Taiwanese investment making China the world leader in the manufacture of desktop and notebook computers, mobile phones, and now computer tablets. Taiwanese companies such as Quanta and Foxconn manufacture these products in China on behalf of the familiar flagship brands—Apple, Dell, Hewlett-Packard, Nokia, Panasonic, and Philips. The Taiwanese "original equipment manufacturers" have also followed the Japanese practice of creating industrial clusters, where plants assembling final products are surrounded by a host of smaller plants turning out components, such as computer motherboards, keyboards, and computer cases.[5]

In the 1990s and 2000s, Taiwanese IT companies transferred virtually their entire labor-intensive production base for IT hardware from Taiwan to the mainland. Between 1995 and 2005, China's share as a production base for Taiwanese hardware production increased from 14 percent to 80 percent, while the share of the Taiwanese home base fell from 72 percent to less than 7 percent.[6] One effect of this mass migration has been to transform Chinese cities, which had been obscure provincial backwaters, into world leaders for the manufacture of particular IT products. Dongguan, in the Pearl River Delta, was formerly a rural township surrounded by rice fields and known for the growing of lychees.

Dongguan is now the world's leading center for the manufacture of desktop computers, with five thousand Taiwanese-owned companies at the center of a production and assembly network turning out power-supply units, printed circuit boards, monitors, and keyboards and also including one thousand indigenous Chinese companies among the component suppliers.[7] In the Yangtze River Delta, Taiwanese investment has transformed Kunshan, formerly a small (by Chinese standards) city surrounded by farmland, into the world's largest manufacturing center for laptop computers. By the early 2000s, there was an expatriate Taiwanese community of fifty thousand in Kunshan, with the Foxconn technology group alone employing sixty thousand workers in the city, including ten thousand student interns.[8]

IN 2010 FOXCONN became the focus of a major scandal when it became known that between January and August 2010, seventeen workers at Foxconn's Chinese plants had attempted suicide, thirteen successfully, and all but one by jumping from the higher floors of the corporation's employee dormitories (the single exception had slit his wrists after failing to jump.)[9] The Foxconn scandal raised leading questions about China's development model. Are the CCP and its mainly offshore Chinese business partners creating a gigantic sweatshop economy in South and Central China, with the connivance of leading US IT corporations such as Apple, HP, and Dell who outsourced most of their computer manufacturing to Taiwanese factories located in China? The answer requires a redefinition of *sweatshop* for the digital age. When I arrived in Hong Kong in the fall of 1966 and began looking at the colony's industries, I encountered sweatshops in their full Dickensian meaning.

One of the unique sounds of Hong Kong, encountered after dark in the industrial back streets of Kowloon, was the din of machines humming away on one story of a building, accompanied by the machinelike noise of mah-jongg counters being shuffled by devotees of the game on a storey below. The fact that the factories were still running after dark was itself significant. At that time the Hong Kong government's "light-touch" regulation of the industrial workplace did not extend to working hours or terms of employment and was barely concerned with health and safety. The most spectacular example of this neglect was also to be found in Kowloon, part of the mainland territory of Hong Kong just opposite Hong Kong Island itself.

This was the Kowloon Walled City, a community with a population of about thirty thousand and with a unique history.[10] When Kowloon was ceded to Britain by China in 1898, the two governments could not agree on who had jurisdiction over the Walled City. None of the succeeding Chinese regimes tried to reestablish a presence in the Walled City, but the British colonial administration, in deference to Chinese sensibilities, had not established a presence there either. For most of the twentieth century, the Walled City was literally ungoverned, unless one counts the triads, the Chinese mafia, and the Walled City's de facto rulers. When I first visited the Walled City in 1966, it was running at full tilt, and to enter it was to enter a twentieth-century version of a medieval city. With no enforced building code, the city's tenements lurched precariously over the narrow streets, leaving them in permanent near darkness. The stench of uncollected garbage and untreated sewage was overwhelming. Drug addicts and teenage prostitutes competed for the tiny parcels of public space. Yet the Walled City had its own economy, the drugs trade apart. It was the ultimate sweatshop, with a network of small workshops turning out textiles, garments, and footwear, operating 24/7. Working conditions there were beyond belief, with sickly, undernourished workers tendering unprotected machines amid the noise, dust, and filth.

Fast-forward now nearly fifty years to Foxconn's and Quanta's giant state-of-the-art electronics plants in the Pearl and Yangtze River Deltas, and there is no trace of this squalor.[11] With their cleanliness, spaciousness, and pastel colors, they have the look of Scandinavian pharmaceutical plants. In some of them employees wear surgical masks and bathroom caps to preserve the purity of their products. But this bland facade coexists with sweatshops in the full sense of the word. Draconian variants of Scientific Management, constantly subject to Business Process Reengineering, have ruled at the factories of the Pearl and Yangtze River Deltas. At "Litton Electronics" (a pseudonym), Ching Kwan Lee of the University of California, Los Angeles, found that "work procedure sheets specifying the minute details of tasks for a particular line position were photocopied and hung over exactly the same seats of the two assembly lines," one in Hong Kong and the other in Shenzen.[12]

In a report on working conditions at Foxconn's Chinese plants, linked to the 2010 suicides, a coalition of Hong Kong scholars and students showed that Foxconn operated a simplified version of the Toyota system labeled "management by stress" by Mike Parker and Jane Slaughter in their book of the same name.[13] On the line at Toyota each worker has access to the "andon cord," linked to an overhead light. When the line is running smoothly, the lights above the line flash green. When the line speed accelerates to the point where a worker can no longer cope, he or she must pull the andon cord, and the overhead light flashes red. The line then stops, and scientific managers descend and simplify the worker's "time and motion" routine to reduce it by a second or two. Then the line resumes at a faster pace.

At Foxconn there were no lights or cords, but "if workers can finish their quota the target will be increased day by day until the capacity of the workers is maximised." In the constant battle to in-crease productivity, managers relied on the classic Taylorist meth-odology of breaking down tasks "into more precise and tedious steps" so that they can be speeded up "and production targets keep surging."[14]

Presiding over Foxconn has been CEO Terry Gou, the near-est the Chinese corporate world has gotten to Chairman Mao of the Cultural Revolution. It was Gou who speculated that Foxconn workers who committed suicide might be motivated by the hope that their survivors would receive financial compensation from Foxconn, thus the letter Foxconn employees were obliged to sign, forswearing such compensation. *Businessweek* has published say-ings of Gou such as "A harsh environment is a good thing," "Hungry people have especially clear minds," and "Outside the laboratory there is no high technology, only execution of discipline." At Fox-conn's Hangzhou plant, a worker who had forgotten to fix a screw on a mobile phone was ordered to write out quotations from Chair-man Gou three hundred times. Also in the manner of the Cultural Revolution, erring workers could be made to perform self-criticism in front of their colleagues at the end of the work shift.[15]

Foxconn, like Amazon, has been willing to push its workforce to the limit, employing the most advanced technologies to do it. But the Chinese workforce is subject to pressures that have no parallel at Amazon or indeed in the United States and are bound up with China's status as a still overwhelmingly agrarian economy where the citizen is subject to the arbitrary powers of the state in ways having no parallel in the West. These pressures further weaken a workforce already undermined by the panoptic controls of mass production and of Business Process Reengineering.

The gigantic Chinese workforce that has manned the factories of the Pearl and Yangtze River Deltas is, as we have seen, composed overwhelmingly of migrant labor moving from China's villages to its cities. In a survey conducted in 2007, the All-China Federation of Trade Unions estimated the strength of the migrant Chinese labor force at 120 million, constituting 64 percent of all those in industrial employment.[16] With the system of household registration still in force in China, the great majority of these migrant workers were registered at their home villages and had to apply for temporary work permits to live and work in the cities. Under the law they remain temporary residents of the cities however long their stay there, and the validity of their temporary permits is linked to their employment. If they lose their jobs and cannot find another, they must go back to their villages.[17]

Their status and bargaining power are further undermined by a practice that has all too many parallels in the United States. Most migrant workers are not only temporary residents of the cities where they live, but also temporary, contingent employees of the companies they work for. Known in China as "dispatch workers," they are employed by hiring companies that send them "to manufacturers in need of highly flexible and highly exploitable workers." With an estimated strength of 270 million, the dispatch workforce is more than twice the size of the migrant workforce and comprises urban workers employed in the service as well as manufacturing economies, as well as employees working for indigenous Chinese companies.[18] As in the United States, these are "just in time" employees who can be easily hired and fired according to short-term fluctuations of demand and who enjoy even less security of employment than coworkers directly employed by their businesses.

In her research Ching Kwan Lee has gathered and published the testimony of employees working at foreign-owned manufacturers who had witnessed how coworkers, like the seventeen workers at Foxconn, had been driven beyond the limit by the relentless demands of production. A line supervisor at one of the world's largest hard-disk manufacturers described how seven young women on the line broke down and had to be committed to mental hospitals:

> The girls thought it was a curse in the factory. But I think it's because of the indescribable stress at work. Managers were ruthless and reprimanded for the most minor mistakes. You got scolded, humiliated and fined for a loosened screw, or dropping something on the floor. On the shop floor, foremen always threatened to "deduct your 107." That's the amount of monthly bonus. Any minor mistake, like being late for a few minutes or taking a day of sick leave, can cost us 107 RMB [about $16]. Some young girls did not know how to deal with this kind of abuse and they just took it all inside themselves. You can see the pain in their deadly silence. At some point they could take it no more and lost their minds.[19]

THIS CHAPTER SO FAR has focused on the "horizontal" processes of mass production and Business Process Reengineering, applied in a Chinese setting to the manufacturing processes of giant plants now globally dominant in the mass production of electronic goods. But the American model of digital capitalism also includes the second, vertical, dimension of Corporate Panoptics, the top-down systems of control that link the horizontal processes of BPR to vertical panoptic information systems at the disposal of top managers. These empower managers to set multiple targets for corporate and employee performance at all levels of the business, with managers then relying on real-time monitoring to establish whether targets are being met.

Although there is a scholarly literature on the introduction of
CBSs in China, most of it is written from a technical, managerial
perspective and does not raise wider issues of ethics and political
economy.[20] The research of Kimberly Chong of the London School
of Economics is rare in doing this. For her PhD thesis, Chong did
fieldwork at one of the three China offices of "Systeo" (a pseudonym), a leading US management consultancy specializing in IT.[21]
The office was located on Dalian island, off the North China coast.
The main task of Systeo's consultants on Dalian, and at their Beijing
office where Chong also spent time, was to apply SAP's Computer
Business Systems, including Corporate Panoptics, to Chinese state-
owned enterprises (SOEs) either recently privatized or being read-
ied for privatization.

This in itself was significant. Systeo's task as consultant was not
to work out a customized "grand strategy" for each of its SOE clients,
but simply to act as glorified software engineers in grafting SAP's
complex and high-priced systems onto what were often gigantic
businesses.[22] Also significant was the acceptance by all the parties
concerned—Systeo, the CCP, and the state-owned enterprises—that
SAP systems were templates of corporate modernity that had to be
present and operational if an SOE was to be successfully privatized
and launched on the international markets.

Once Systeo's consultants came into actual physical contact
with the SOEs, they often found that they were in poor shape.
This was hardly surprising after a sixty-year history in which the
SOEs had endured Soviet-style planning, the Cultural Revolution,
and years of uncertainty and neglect during the early phases of the
post-Mao reforms. One of China's largest SOEs, PetroChina, was
so overmanned that it laid off *1 million* employees in preparation
for privatization.[23] Chong gives an illuminating and often hilarious

account of what the Systeo consultants found once they got inside the SOEs. Chong herself visited the premises of one of Systeo's main clients, "China Utility" (a pseudonym), and found "not an office but a large unheated hall of a former socialist work unit which had previously been used as a table tennis facility."

It was winter in Beijing, and the temperature was approaching 14 degrees Fahrenheit. Consultants were wearing down jackets inside and had sealed the windows "using duct tape in a vain attempt to keep out the cold." On the IT side there were "eclectic collections of computers, some desktops, some laptops, Dells, Compaqs, Lenovos and HPs, all thrown together upon rows of grey desks." At the beginning of the working day, Systeo consultants would start looking over the shoulders of the China Utility employees, guiding them in their use of the SAP software. Some employees followed instructions, but others preferred to play computer games: "This was done openly with consultants seemingly indifferent to the cluster of end users who would leave their seats to watch the antics of another China Utility employee immersed in a game of *Formula Racing*."[24]

Senior management at Systeo became worried that their consultants, who could be posted for as long as three years with SOE clients such as China Utility, might themselves become infected with these bad "old ways" of the former socialist workplace. The intention of at least the senior management at Systeo was to banish these old ways for good, so that middle and lower managers and front-line workers at the Chinese SOEs would work with the speed and efficiency of workers on the line at Foxconn or Quanta.

But there was another sense in which senior managers at Chinese SOEs would under this regime find themselves very much at home with the systems of Corporate Panoptics that their American-trained consultants wanted to introduce. This familiarity pivots on the practices of Kaplan and Norton's Balanced Scorecard, and especially of "cascading," which were embedded within SAP systems.

It is fitting that one of the seminal documents on "cascading" is jointly authored by David Norton, the junior member of the Kaplan-Norton team that has marketed the Balanced Scorecard from the 1990s onward, and Juergen Daum, at the time a senior system designer at SAP and now an independent consultant. The title of the document is *SAP Strategic Enterprise Management, Translating Strategy into Action: The Balanced Scorecard.*[25] Norton and Daum's definition of *cascading* is central to an understanding of what Systeo and its competitors were trying to do with the Chinese SOEs and why it is not out of line to view their activities as a variant of old Soviet planning methods but in an advanced, digital form.

Indeed, I have often thought that if Mikhail Gorbachev could have held on for just a few more years, the whole apparatus of Soviet planning, from the central planning agency, USSR Gosplan, down to the humblest machine-tool plant in Omsk, might have been given a reprieve at the hands of teams of US consultants armed with SAP and IBM systems and with the practices of the Balanced Scorecard, and of "cascading" in particular, embedded within them. Here in full is Norton and Daum's definition of *cascading* in *SAP Strategic Enterprise Management:*

> The Balanced Scorecard is the linchpin between change initiated by a small number of people at the top and executed by a large number of people at the bottom. The scorecard is a way to translate the strategy at the top so it can be made operational at the bottom. By cascading the scorecard from the "boardroom to the backroom" a powerful new framework has emerged to create strategy-based performance-management systems. Individuals

can implement strategy only when they clearly understand it and realize how they can contribute to its achievement. Traditional human resource systems and processes play an essential role in enabling this transition.[26]

The Soviet planning system embodied primitive, dysfunctional versions of cascading. USSR Gosplan's targets for the global production of every Soviet industry was cascaded down into a host of subtargets that the tens of thousands of individual enterprises had to attain if the industry as a whole was to meet its global target. In an inspired work of comparative sociology, "A Sovietological View of Modern Britain," British sociologist Ronald Amann shows that there was a striking similarity between this cascading of the Soviet planning system during the post-Stalin era and that of the bureaucratic hierarchy set up by successive British governments to manage the United Kingdom's academic production regime, as described in Chapter 4. Amann chose the Soviet machine-tool industry as his Soviet case study.[27]

At the beginning of his piece, Amann includes an organizational chart, matching each Soviet entity with a British counterpart. So USSR Gosplan, the supreme Soviet planning agency, matched with the UK Treasury; the Soviet Machine Building Ministry matched with the Higher Education Funding Council for England, or HEFCE, the UK Treasury's agent in running the academic production regime; Uralmashzavod, the giant Soviet engineering combine, matched with the corporate university; and the combine's Machine Tool Division matched with a university school or division, such as Oxford's Division of the Humanities. In the British system the global targets for research output set by the UK Treasury were cascaded down via HEFCE into a host of subtargets that university divisions and departments had to attain in order to avoid a loss of funding.[28]

Chong's word for *cascading* is *translation,* which is an apt synonym for a process consisting of "translating high level managerial strategy into concrete actions at the lower levels."[29] The big difference between the Soviet and contemporary corporate versions of cascading is that the information systems of the former were primitive to nonexistent, whereas those of the latter are advanced and elaborate. With the unveiling of its five-year plans, USSR Gosplan and indeed its Chinese equivalent, the State Planning Commission, could proclaim its multiple cascading targets, but it had no reliable way of knowing whether the targets bore any relation to what the thousands of subordinate enterprises could produce and whether once the targets had been proclaimed, they were being attained.

The outcome was, in Amann's words, "a very system-specific form of human capital adoption" that got by with "bluffing deception, bullying and bribing." The system also produced "goods that nobody actually wanted because in place of real customers, enterprises faced crude plan targets, often expressed in terms of volume or weight or materials used"—a description that also fits the avalanche of mostly pointless research produced by the United Kingdom's academic production system whose volume, as Amann points out, was so bulky that it has had to be stored in a giant aircraft hangar located near HEFCE's headquarters at Bristol, England.

But why should "cascading" or "translation" work any better with Systeo's clients among the Chinese SOEs, or indeed with its clients in corporate America? Here we are back in the realm of the key performance indicators, or KPIs, the central nervous system of the whole Balanced Scorecard regime. What in the Soviet era were known as targets are now KPIs, though every KPI comes with its own target attached. As we saw in Chapter 1, there is no limit to the number of KPIs that system designers can attach to business processes at all levels, including the work of employees or teams of employees engaged in a process. But however numerous the KPIs, they can be lodged in the system's database and can be accessed immediately. Also accessible in real time is information on whether KPIs are being met by business actors, from corporate divisions to single employees. Another feature of the system, unavailable to the Soviet planners, is the power to change rapidly the whole hierarchy of KPIs in light of changing economic circumstances, to increase them if conditions improve or reduce them if conditions deteriorate.

Veterans of the Chinese SOEs, with plans and targeting in their blood, will now find themselves disposing of powers that under the ancien régime were reserved for the State Planning Commission and that in their number and variety go far beyond the wildest dreams of their managerial predecessors of the Mao era. The research of this book suggests that these Chinese KPI regimes will be more successful in businesses where production pivots on the manufacture and movement of goods, the Chinese equivalents of Walmart and Amazon, and less successful in industries involving complex interactions between human agents, as in the Chinese health care and banking sectors.

But in a Chinese setting, these questions of economic performance cannot be looked at in isolation from Corporate Panoptics as a phenomenon of political economy. The transfer of CP from West to East brings to Chinese managerial and clerical offices the disciplines that mass production and Business Process Reengineering have brought to the factory floor. This upward mobility fits in well with the CCP's outsourcing of workplace control to its management allies, extending to these higher-echelon control regimes infinitely more powerful than those of the old party committee of the Mao era. Are we then witnessing the birth of a new authoritarianism of the digital age, with the existence of two parallel and mutually reinforcing hierarchies of control, Party and Business, each strengthened by the coercive powers of the other, and both relying on the fruits of economic success to pacify a citizenry subject to the coercive powers of both. And how does the West counter the global appeal of an economic model that seems to demonstrate that economic success can be achieved without the inconveniences and complications of liberal democracy?

10

ANY WAY OUT?

THE "CLOSED WORLD" IS A STATE OF MIND AND A WAY OF THINKING that straddles the military and civilian worlds of CBSs, bringing a dehumanized rationality to the workings of the civilian economy. With an authoritarianism bound up with its part-military origins, the civilian "closed world" dehumanizes us by turning us into abstracted electronic and statistical entities subject to the system's science-based rationality. Although this rationality seems to take on an impersonal, abstracted form, all the system's rules and commands in fact have human origins in the superior expertise of the technical, managerial elite whose wisdom is baked into the system. This is what "dumbing down" means in the early twenty-first century.

This dumbing down is also joined at the hip to another practice, one that was a dominant force in American capitalism during the twentieth century and shows every sign of extending its dominance to the twenty-first. Scientific Management has been a looming presence throughout this book, as it is a looming presence in the contemporary US economy. But for the theorists and practitioners of the CBS world it is a Great Unmentionable, and by keeping it out of their texts and out of public sight they have brought off a distortion of American business history worthy of *Pravda* itself. The systems they have created are essentially latter-day vehicles of Scientific Management, vastly empowered by information technology.

In concealing this Taylorist ancestry, the theorists have benefited from the popular identification of Scientific Management with the rudimentary work practices of Taylor's own time—the shoveling of sand and clay in the backyards of the Bethlehem Steel company in the 1890s or the mindless repetitions of the Fordist assembly line two decades later. But the core principle of Scientific Management that shaped these processes is no less applicable to the digital processes of our own time, however distant in place, time, and character they may be from their primitive forebears.

The central principle of Scientific Management has been and remains the separation of the detailed planning of work from its execution—the definition of the practice that Harold Leavitt and Thomas Whisler relied on in their landmark 1958 piece for the *Harvard Business Review,* "Management in the 1980s." From this unequal division of labor stem many of the inequalities of skill, power, and income noted in embryonic form by Leavitt and Whisler more than fifty years ago and rampant in the US economy of today. These inequalities separate those who control and create the systems—the managerial and technocratic elite inhabiting Leavitt and Whisler's football perched at the pinnacle of their workplace pyramid—from those who are subject to the system's orders and have to follow them.

These inequalities are rampant because the information technologies embedded in CBSs are being applied to ever-widening segments of the service economy, most notably to core services, complex dealings between human agents, as in health care, financial services, human resource management, and customer relations management, as I have tried to show in this book. But there are aspects of this white-collar industrialism that differentiate it from all its industrial forebears and can be easily overlooked. With the assembly line, the machine shop, and the Amazon fulfillment center, the disciplines of Scientific Management link two parties—on one side the expert engineers who as scientific managers determine exactly how shop-floor routines should be performed and on the other the front-line workers who follow orders.

But with core services Scientific Management brings in a third party—ourselves as customers, clients, students, householders, and patients. On the line at Ford or Amazon the objects of work are inanimate and inert, goods or components that workers must machine, join together, package, or shift. But the objects of core services are very much alive, because they are us, most notoriously in human resource management, customer relations management, financial services, and health care. Here Scientific Management has to take the form of iconic, authoritative databases that, embedded in expert systems, set out to cover all the human contingencies that front-line workers have to deal with, thus the preplanned digital script that I encountered in my dealings with the Toshiba customer service agent.

Without such guidance from above, employees would have to exercise their own judgment and skill in resolving the problems, often fraught with human complexity, that they encounter every day in their working lives when they have to deal with us as counterparties. These encounters are built around what social anthropologist Lucy Suchman, perhaps the most cogent and influential critic of CBS orthodoxies, has called "situated actions"—actions rooted in the infinite richness and diversity of human life itself. But the acknowledgment and accommodation of such diversity, and the devolution to employees of the power and skill to deal with it, are incompatible with Scientific Management as a practice of white-collar industrialism.[1]

When applied to core services Scientific Management deals instead in "unsituated actions," meaning that the chief task of the employee, whether as a physician or a call center agent, is not to apply her own judgment and skill to the complex situations she encounters. Instead, her task is to work out which predetermined category a patient, client, householder, or customer belongs to, so that the appropriate treatment, reward, or advice already worked out by scientific managers can be applied. But these rudimentary classifications drag us as customers into the maw of Scientific Management, no less than they do the employees we are dealing with. Just as the core workplace relationship between the managers and the managed denies employees the scope to develop their talents, so the further relationship of the employees with us as their counterparties too often deals in an attenuated version of ourselves that, hedged about by the ever-present constraints of time, can too easily end up as abstract representations on a digital screen. For those with the financial wherewithal, the concierge economy with all its enticements is an escape from this rushed, dehumanized world.

OF ALL THE case histories we have examined here, perhaps the most chilling are those situated well beyond the strictly business world and even beyond the gaze of tightly networked computers. These are the theories and practices of the emotional labor experts of HRM, which intrude upon our innermost beings, and the efforts to impose an industrial production regime at a great university such as Oxford. These ventures show that the vandalism of the CBS world knows no limits, and there is no corner of our lives that is beyond the reach of *process*. They also show that you do not need an elaborate network of computers and software to make the systems work. You can do it with nondigital methods, as the practices of the emotional labor theorists demonstrate.

In this CBS world concepts such as empowerment and skill no longer mean what they once did. To be skilled and empowered is to be in a state of perfect, frictionless harmony with the system, in perfect conformity with its rules and commands. Because experience and wisdom reside in the system and not in those who use it, the experience that users accumulate over time does not make them any more valuable to the system. Indeed, the contrary is true, because older workers may become wedded to past practices of the system that are now obsolete. These veterans can and should be fired and replaced by younger workers who can be paid less and have no crusty attachments to past practices.

But let's pause a moment and play devil's advocate. What if the wisdom baked into the systems *is* superior; if the reengineers *are* better qualified to streamline business processes than the workers who operate them; if the all-seeing eye of corporate panoptics *can* nose out the pockets of process inefficiency and remove them; if labor unions, with their attachment to the obsolete, are obstacles to progress; and if a sentimental attachment to a workplace humanism simply opens up the field for the hard-nosed Chinese with no such scruples? The moral here is that the case against the new digital industrialism needs in the first instance to be an economic case, because the ethical divorced from the economic lacks traction. But the economic dragged down by the unethical is another matter.

So perhaps the most powerful argument against the new digital industrialism is the one outlined in the introduction to this book. By relying on information technology both to accelerate the processes of business output and to diminish the role of labor in production, along with its earning power, the new digital industrialism has overlooked the identity between producers and consumers, ignoring the wisdom of Henry Ford when he introduced the five-dollar day at his Detroit plants in 1914. Ford saw that his workers needed to be well paid in order to afford the Model Ts rolling off the line at the Highland Park plant. In contrast, the earnings malaise of today's US producers is now spilling over into their lives as consumers, after years of putting off the evil hour with recourse to debt. With American consumers providing 75 percent of the economy's final demand, this is a serious blow to the economy's growth prospects, a leading cause of the weak recovery from the Great Recession, and a reason the Fed has had to keep interest rates so low and for so long.

What is to be done? Before looking for answers, it is worth taking stock of the headwinds most Americans face, judged by the statistics for the long-term stagnation or decline of their real earnings: first, their employment in workplaces that do not make full use of their skills and subject them to intrusive systems of monitoring and control; second, the stagnation or shrinkage of their real earnings, related directly to this deskilling; third, their need to shoulder increasing health care and pension costs, dumped on them by employers; and fourth, the growing insecurity of the workplace, linked to outsourcing, globalization, and a corporate readiness to have early recourse to layoffs.[2] These are not the acts of a corporate leadership that values the skills and loyalty of its workforce and wants to strengthen these ties over time. These are indeed the claims of countless corporate mission statements, but the record reveals a preference for harsh cost-cutting strategies in which high employee turnover and high employee cynicism can be offset by system expertise and with the system's control mechanisms ensuring that employees act as the systems prescribe.

It is not hard to see what now needs to be done: the creation of higher-paying, higher-skilled jobs, with the component CBS technologies used to supplement rather than replace employee expertise, backed by effective institutions of education and training and with good performance recognized and rewarded. The sector that best exemplifies these qualities is the IT sector itself. But the relation between the IT industries and the CBS control regime they have spawned is highly ambivalent. In 2000 I visited a Silicon Valley software start-up called Clarify, Inc., of San Jose. Like many such start-ups, it was acquired in 2001 by the Israeli software company Amdocs. The setup at Clarify exemplified everything that was hip, fluid, and creative about Silicon Valley workplace culture: no discernible hierarchies, an office that had the feel of an upmarket California living room, the CEO dressed in the Steve Jobs uniform of sneakers, jeans, and a T-shirt. The ambiguity arose when I asked the CEO about his product, which was workflow management software. He said, "Our systems are paradise for managers. They can know and control everything. I'd like to be a manager in a company using our software." Outward signs of hipness did not necessarily translate into an inner hipness. The role of the products these hipsters created was to repress among their clients the very qualities that underwrote their own success.

There are case histories both in the United States and in Europe where alternative, employee-friendly cultures have taken root, usually for reasons specific to a particular location or company and not easily replicated elsewhere: the culture of German codetermination and labor-management partnership, taking shape at the Chemnitz machine-tool plant in 1992; the Scandinavian tradition of employee participation in software system design, especially strong in Norway and Denmark; the Mondragon cooperatives in the Basque country of northern Spain, employing more than eighty thousand workers and dominating the regional manufacturing economy; and the John Lewis Partnership in the United Kingdom, employee owned and the best high-quality retail chain in the country.[3]

In the United States there are exceptional companies like Lincoln Electric, with generous profit sharing for employees and a policy of "no layoffs" even in times of recession, a commitment honored by the company since 1948; the employee-friendly expert systems at Xerox, a legacy of Xerox's progressive Palo Alto Research Center; companies like Microsoft who buck the trend and provide good customer service; or a foundation such as the Regenstrief Institute of Indianapolis, which has helped put the medical records of Indianapolis residents online, a system built successfully from the medical grassroots upward with the support and participation of physicians, nurses, and patients.[4]

Yet it would be delusional to think that, in the United States, the domain of these alternative work cultures will expand spontaneously by virtue of their ethical strengths and their proven record of success in the marketplace. They come up against the hard armor of corporate power, with CBSs as corporate marine guards and with white-collar industrialism yielding a distribution of income exceedingly favorable to the managerial elite. John Bakija of Williams College and Bradley T. Heim of Indiana University have shown that this elite dominates the top 0.1 percent of taxpayers who have been the biggest winners from the growth of American inequality. In 2004 69.2 percent of these winners were by profession "executives, managers, and supervisors in financial and non-financial companies."[5]

If a clear majority of Americans are losing out in today's economy, as they are, the political task is to create a dominant coalition from among them that would include low-income minorities and whites of the Walmart and Amazon world, middle managers and middle administrators whose real incomes have been steadily eroding, and even nonelite professionals of the nonconcierge economy suffering the same fate. The political debate is central, and it should be very much part of this debate that the progressive critique of the economy include the issues of white-collar industrialism discussed here.

The progressive response to the harshness of nineteenth-century capitalism was fueled by a growing awareness of what was going on behind factory walls. CBSs are by comparison invisible, and they benefit from this obscurity. This needs to end, and this books is a modest step in that direction. Yet there are grounds for optimism. The future contours of the economic debate are fluid because the future course of the economy itself is fluid. With its failure to reward the majority of Americans, the economy's present course is unsustainable, and as this becomes more and more apparent, volatility will spill over to the public debate and open it up.

In macroeconomics this unsustainability goes beyond the preoccupation with public spending and the public debt, currently the number-one concern in Washington. It is bound up with the difficulty of achieving strong, sustained growth as long as consumer-producers are in eclipse, blunting what was once the economy's most reliable source of demand and making the tasks of deficit reduction immeasurably harder. But the politics of the wounded producer-consumer is a whirlpool of volatility.

The Tea Party's resentment is, for example, promiscuous in its choice of enemies, big business as well as big government, and with

Mitt Romney as archetypal corporate schmoozer scarcely more acceptable to Tea Party militants than the supposedly leftist Obama. Similarly, the willingness of the hard-pressed white working class to vote for procorporate Republicans—so well described by Thomas Frank in *What's the Matter with Kansas?*—still ranks among the most spectacular examples of the Marxian false consciousness of recent times.[6] Obama's success in the 2012 presidential election in blunting the "Kansas effect" in the state of Ohio, critical to his overall victory, and much aided by Mitt Romney's kamikaze politics on the General Motors and Chrysler bailouts, is a sign perhaps that a working-class realignment is under way. Progressives and their union allies should now bend every sinew to keep this going, building on Republican blunders and settling once and for all who the American worker's chief enemies really are.

NOTES

Notes to Introduction: Toward a New Industrial State

1. Data available at the website of Emmanuel Saez of the University of California, Berkeley, who, with Thomas Piketty of the Paris School of Economics, based this research on an examination of the tax returns of the richest Americans, available at the Internal Revenue Service. See http://elsa.berkeley.edu/~saez/TabFig2007.xls.

2. National Center for Educational Statistics, "Economic Outcomes—Table 20-1. Median Annual Earnings of Full Time, Full Year Wage and Salary Workers Ages 25–34 by Educational Attainment, Sex and Race/Ethnicity: Selected Years 1980–2006," US Department of Education, Institute of Education Sciences, http://nces.ed.gov/programs/coe/2008/section2/table.asp?tableID=894, quoted in Jacob S. Hacker and Paul Pierson, *Winner-Take-All Politics: How Washington Made the Rich Richer and Turned Its Back on the Middle Class* (New York: Simon and Schuster, 2011), 36.

3. Michael Hammer, *The Reengineering Revolution* (New York: Harper Business, 1995), xi.

4. Eric Brynjolfsson and Andrew McAfee, "Investing in the IT That Makes a Competitive Difference," *Harvard Business Review,* July 2008.

5. See, for example, Kimberly Chong, "Financialization: An Ethnography of a Global Management Consultancy in Post-Mao China" (D.Phil. diss.,

London School of Economics, 2012). See also the discussion of Chong's work in Chapter 9, "The Chinese Model."

6. *Encyclopaedia Britannica, 1929 Edition*, s.v. "Mass production."

7. Henry Ford, *My Life and Work*, quoted in Stephen Meyer, *The Five Dollar Day: Labor, Management, and Social Control in the Ford Motor Company, 1908–1921* (Albany: SUNY Press, 1981), 21.

8. Alfred D. Chandler, *The Visible Hand: The Managerial Revolution in American Business* (Cambridge, MA: Harvard University Press, 1993).

9. Adam Smith, *The Wealth of Nations*, bk. 1, chap. 1; Karl Marx, *Capital*, esp. vol. 1, pt. 4, chap. 15, "Machinery and Modern Industry"; Horace Arnold, *Ford Methods and the Ford Shops* (Boston: Adament Media, 2005); James P. Womack, Daniel T. Jones, and Daniel Roos, *The Machine That Changed the World: The Story of Lean Production* (New York: Free Press, 2007).

10. See, for example, Peter Drucker, *Post-Capitalist Society* (New York: HarperCollins, 1993).

11. Board of Governors of the Federal Reserve System, press release following meeting of the Federal Open Market Committee, September 21, 2010, www.federalreserve.gov/newsevents/press/monetary/20100921a.htm.

Notes to Chapter 1: Inside the Belly of the Beast

1. See, in particular, Chuck Ballard et al., *Business Performance Management . . . Meets Business Intelligence*, 2004, http://ibm.com/redbooks; Upannee Amnajmongkol et al., *Business Activity Monitoring with WebSphere Business Monitor V6.1*, http://ibm.com/redbooks; Oracle Corporation, *Enterprise Performance Management: Bridge to Fusion Application*, April 2011, www.oracle.com/us/solutions/ent-performance-bi/epm-bridge-to-fusion-apps-wp-359993.pdf; Oracle Corporation, *The Path to EPM (Enterprise Performance Management) Success*, June 2008, www.oracle.com/us/solutions/business-intelligence/064034.pdf; Gerhard Keller and Thomas Teufel, *SAP R/3 Process Oriented Implementation* (Reading, MA: Addison Wesley, 1998); Juergen Daum and David Norton, *SAP Strategic Enterprise Management, Translating Strategy into Action: The Balanced Scorecard*, www.juergendaum.

com/news/sap_sem_wp_bsc.pdf; SAP, *Solutions for Business Activity Monitoring*, www.sap.com/belux/platform/netweaver/pdf/BWP_Business_Activity_Monitoring.pdf; August-Wilhelm Scheer et al., *Corporate Performance Management: ARIS in Practice* (Berlin: Springer, 2005); and August-Wilhelm Scheer et al., *Business Process Automation* (Berlin: Springer, 2004).

2. Amnajmongkol et al., *Business Activity Monitoring*, 4.

3. See, for example, SAP, *mySAP Customer Relationship Management*, 2003, www.sap.com/uk/solutions/sam/pdf/BWP_Automotive.pdf; SAP, *Solutions in Retail: Enterprise Compensation Management with mySAP ERP*, 2004, www.sap.com/uk/solutions/business-suite/erp/pdf/BWP_SID_Enterprise_Compensation.pdf; SAP, *mySAP ERP Human Capital Management, Solution Overview*, www.sap.com/asia/pdf/2011/SAP-ERP_Human_Capital_Management.pdf; SAP, *Intelligent Workforce Management*, 2004, www.sap.com/uk/solutions/business-suite/crm/pdf/BWP_Retail_Intel_Workforce_Mgmt.pdf; and SAP for Higher Education and Research, http://scn.sap.com/community/higher-education-and-research.

4. IBM, *WebSphere Business Monitor, Version 6.0*, 2005, www-01.ibm.com/common/ssi/cgi-bin/ssialias?infotype=an&subtype=ca&html fid=897/ENUS205-227&appname=isource, 2.

5. See Stephanie Smith and Stephanie Bancharo, "Putting a Price on Professors," *Wall Street Journal*, October 22, 2010.

6. See the discussion of software for human resource management in Chapter 4.

7. IBM, Executive Brief, *Business Activity Management: Your Window of Opportunity for Better Business Operations*, 2003, ftp://service.boulder.ibm.com/eserver/zseries/audio/pdfs/WBIMonitorbrief, 9, 16.

8. IBM, *WebSphere Business Monitor*, "Frequently Asked Questions," www-01.ibm.com/software/integration/wbimonitor/faq/, 2.

9. Scheer et al., *Corporate Performance Management*, pt. 1, chap. 1; Helga Hess, *From Corporate Strategy to Process Performance: What Comes After Business Intelligence?* (Berlin: Springer, 2006), 28.

10. Scheer et al., *Corporate Performance Management*, pt. 1, chap. 2; Andreas Kronz, *Managing of Process Key Performance Indicators as Part of the ARIS Methodology* (Berlin: Springer, 2006), 36–39.

11. IBM Redbooks, *Lotus Domino Domain Monitoring*, www.redbooks. ibm.com/abstracts/redp4089.html.

12. Scheer et al., *Corporate Performance Management*, 13, 47–48.

13. Ballard et al., *Business Performance Management*, 33–34; Amnajmongkol et al., *Business Activity Monitoring*, 50.

14. IBM, *Websphere Business Monitor, Version 6.0, Highlights*, ftp://ftp .software.ibm.com/software/emea/de/websphere/G224-7530-00.pdf, 2.

15. Quoted in IBM, *Business Activity Management: Your Window of Opportunity for Better Business Operations*, ftp://service.boulder.ibm.com/ eserver/zseries/audio/pdfs/WBIMonitorbrief.pdf, 5.

16. Cognos Corporation, *The Evolution of the CPM System*, www.bicon cepts.at/fileadmin/user_upload/downloads/Cognos_CPM.pdf, 4. Ash was at the time CEO of Cognos, which was acquired by IBM in 2008.

17. IBM, *WebSphere Business Monitor*, "Frequently Asked Questions," www-01.ibm.com/software/integration/wbimonitor/faq/.

18. Scheer et al., *Corporate Performance Management*, 45, 2; and especially Markus von den Driesch and Tobias Blickle, *Operational, Tool-Supported Corporate Performance Management with the ARIS Process Performance Manager* (Berlin: Springer, 2006), 45–64.

Notes to Chapter 2: Walmart and Amazon

1. www.mckinsey.com/insights/mgi/research/productivity_competitive ness_and_growth/us_productivity_growth_1995-2000.

2. Ibid.

3. Ellen Rose, "The Quality of Work at WalMart," paper presented at the conference "WalMart: Template for 21st Century Capitalism," University of California, Santa Barbara, April 12, 2004. See also Simon Head, "Inside the Leviathan," *The New York Review of Books*, December 16, 2004, available at http://nybooks.com/articles/archives/2004/dec/16/inside-the-leviathan-2/.

4. Barbara Ehrenreich, *Nickel and Dimed: On (Not) Getting by in America* (New York: Metropolitan Books), chap. 3.

5. John Marshall, "The High Price of Low Cost: The View from the Other Side of Walmart's 'Productivity Loop,'" www.google.co.uk/url?sa =t&rct=j&q=&esrc=s&frm=1&source=web&cd=1&ved=0CC0QFjAA

&url=http%3A%2F%2Fmakingchangeatwalmart.org%2Ffiles%2F2011
%2F10%2FThe-High-Price-of-Low-Cost.pdf&ei=km5MUuWlL-GN0AWw
8oCABQ&usg=AFQjCNEif_e_-pTlzDybmGTYCpUtQRqKgw&bvm=bv
.53371865,d.Yms, 10.

6. Ibid.

7. Ibid., 8.

8. Ibid.

9. Ibid., 22.

10. John Marshall, "Wal-Mart's Labor Problem: Limits to the Low-Road Business Model," http://makingchangeatwalmart.org/files/2012/10/WalmartsLaborProblem.pdf, 10.

11. "WalMart: A Manager's Toolbox to Remaining Union Free (Confidential)," www.ufcw.ca/Theme/UFCW/files/ManagersToolbox.pdf.

12. See Steve Greenhouse, "Wal-Mart Workers Try the Nonunion Route," *New York Times,* June 15, 2011; Greenhouse, "Wal-Mart Workers Stage a Walkout in California," *New York Times,* October 4, 2012; Ricardo Lopez, "Walkout Protests Wal-Mart Warehouse Working Conditions," *Los Angeles Times,* July 24, 2013.

13. Marshall, *High Price of Low Cost,* 10–11.

14. "The Best Performing CEOs in the World," *Harvard Business Review,* http://hbr.org/2013/01/the-best-performing-ceos-in-the-world. See also CNN, "World's Most Admired Companies," http://money.cnn.com/magazines/fortune/most-admired.

15. Mark A. Onetto, address at the Darden School of Business, University of Virginia, October 20, 2009, www.youtube.com/watch?v=Foy1FTBjHK4.

16. Ibid., 8–9; references are to minutes into the lecture, not pages.

17. Ibid., minute 10.

18. "Was War Los in Leipzig," Ver.Di Blog, December 21, 2012, www.amazon-verdi.de/2012/12/04/was-war-los-in-leipzig/, 2.

19. Sarah O'Connor, "Amazon Unpacked," *Financial Times,* February 8, 2013, www.ft.com/cms/s/2/ed6a985c-70bd-11e2-85d0-00144feab49a.html#slide0.

20. Spencer Soper, "Inside Amazon's Warehouse: Lehigh Valley Workers Tell of Brutal Heat, Dizzying Pace at Online Retailer," *Allentown (PA) Morning Call,* September 18, 2011, 6–7, http://articles.mcall.com/2011

-09-18/news/mc-allentown-amazon-complaints-20110917_1_warehouse
-workers-heat-stress-brutal-heat.

21. "Hinter den Kulissen: Das Tagebuch einer Amazon-Packerin," *Frankfurter Allgemeine Zeitung,* February 21, 2013, www.faz.net/aktu ell/wirtschaft/unternehmen/hinter-den-kulissen-das-tagebuch-einer -amazon-packerin-12089481.html.

22. Ibid., 5.

23. Ibid.

24. Onetto, address at the Darden School of Business, 37.

25. Soper, "Inside Amazon's Warehouse," 7.

26. Ibid., 9, 10.

27. Ibid., 9.

28. Vanessa Veselka, "In the Wake of Protest: One Woman's Attempt to Unionize Amazon," *Atlantic Monthly,* December 12, 2011, www.theatlantic. com/technology/archive/2011/12/in-the-wake-of-protest-one-womans -attempt-to-unionize-amazon/249853/, 8.

29. "Das Tagebuch einer Amazon-Packerin," *Frankfurter Allgemeine Zeitung,* February 21, 2013, 6.

30. "Amazon Unpacked," *Financial Times,* February 8, 2013, 5.

31. ARD, "Ausgeliefert, Leiharbeiter Bei Amazon," February 13, 2013, www.ardmediathek.de/das-erste/reportage-dokumentation/ausgeliefert -leiharbeiter-bei-amazon?documentId=13402260.

32. Soper, "Inside Amazon's Warehouse."

33. Ibid.

34. See "Von der Leyen Verlangt Auklarung bei Amazon," *Spiegel Online,* February 16, 2013, www.spiegel.de/wirtschaft/unternehmen/schikanierte-ar beiter-von-der-leyen-verlangt-aufklaerung-bei-amazon-a-883863.html.

35. "Amazon-Chef will Mehr Betriebsrate," *Spiegel Online,* February 21, 2013, www.spiegel.de/wirtschaft/unternehmen/amazon-chef-ralf-kleber -zu-ausbeutungs-und-kartellvorwuerfen-a-884793.html-ralf-kleber -zu-ausbeutungs-und-kartellvorwuerfen-a-884793.html.

Notes to Chapter 3: A Future for the Middle Class

1. See S. J. Prais and Karin Wagner, "Some Practical Aspects of Human Investment: Training Standards in Five Occupations in Britain

and Germany," *National Institute Economic Review* 105, no. 46 (1983), http://ner.sagepub.com/content/105/1/46.full.pdf; and Pepper D. Culpepper and David Finegold, *The German Skills Machine: Sustaining Comparative Advantage in a Global Economy* (New York: Bergahan Books, 2000).

2. Jack Whalen and Erik Vinkhuyzen, "Expert Systems in (Inter)Action: Diagnosing Document Machine Problems over the Telephone," in *Workplace Studies: Recovering Work Practice and Information Systems Design*, edited by Christian Heath, Jon Hindmarsh, and Paul Luff (Cambridge: Cambridge University Press, 2000), 92–140. See also Simon Head, *The New Ruthless Economy: Work and Power in the Digital Age* (Oxford: Oxford University Press, 2005), chaps. 5–6.

3. See particularly Obama's speech at Osawatomie, Kansas, December 6, 2011, www.whitehouse.gov/the-press-office/2011/12/06/remarks -president-economy-osawatomie-kansas; and his speech at Galesburg, Illinois, July 24, 2013, www.whitehouse.gov/the-press-office/2013/07/24/ remarks-president-economy-knox-college-galesburg-il.

4. Statistics available at the United States Department of Labor, Bureau of Labor Statistics, "Manufacturing Employment: Historic Data," www.bls. gov.

5. United National Commodity Trade Data Base, www.indexmundi. com/trade/exports/?country=de.

6. See Obama's speech at Osawatomie, Kansas.

7. See Tamar Lewin, "Community Colleges Cutting Back on Open Access," *New York Times*, June 23, 2010, www.nytimes.com/2010/06/24 /education/24community.html?pagewanted=all.

Notes to Chapter 4: Managing the Human Resource

1. Thomas W. Malone, Kevin Crowston, and George A. Herman, eds., *Organizing Business Knowledge: The MIT Process Handbook* (Cambridge: MIT Press, 2003), esp. 31–37, 389–402, 408–421.

2. Ibid., 394.

3. Ibid., 408–417.

4. Barbara Ehrenreich, *Bait and Switch: The (Futile) Pursuit of the American Dream* (New York: Metropolitan Books, 2005), 32–34, 36–39.

5. Harold Pinter, *Party Time and the New World Order* (New York: Grove Press, 1993).

6. Ibid., 37.

7. Phil Baty, "Lammy Demands Further and Faster Progress Towards Economic Aspect," *Times Higher Education Supplement,* September 10, 2009. See also Simon Head, "The Grim Threat to British Universities," *The New York Review of Books*, January 13, 2011, available at http://nybooks.com/articles /archives/2011/jan/13/grim-threat-british-universities/.

8. For HEFCE's Strategy Statement, see www.hefce.ac.uk/about/how weoperate/strategystatement/.

9. John F. Allen, "Research and How to Promote It in a University," *Future Medicinal Chemistry* 2, no. 1 (2010), jfallen.org/publications.

Notes to Chapter 5: The Case of Goldman Sachs

1. Joseph Schumpeter, *Capitalism, Socialism, and Democracy* (New York and London: Routledge, 2010), 117–118.

2. Simon Johnson and James Kwak, *Thirteen Bankers* (New York: Pantheon, 2010), esp. chaps. 4–5.

3. Nassim Taleb, "Against Value-at-Risk: Nassim Taleb Replies to Philippe Jorion," www.fooledbyrandomness.com/jorion.html. See also Philippe Jorion, "In Defense of VAR," 1997, http://merage.uci.edu /~jorion/oc/ntalib2.html.

4. US Senate, Permanent Subcommittee on Investigations of the Committee on Homeland Security and Governmental Affairs, "Wall Street and the Financial Crisis: Anatomy of a Financial Collapse," sec. C, "Failing to Manage Conflicts of Interest: Case Study of Goldman Sachs," at www.hs gac.senate.gov//imo/media/doc/Financial_Crisis/FinancialCrisisReport. pdf?attempt=2, 376–639.

5. US Senate, Permanent Subcommittee on Investigations, "Wall Street and the Financial Crisis: The Role of the Investment Banks," Hearings, April 27, 2010, www.google.co.uk/url?sa=t&rct=j&q=&esrc=s&frm=1&-source=web&cd=3&ved=0CD0QFjAC&url=http%3A%2F%2Fwww .hsgac.senate.gov%2Fdownload%2F%3Fid%3D0adb7a91-468b-413b -ad52-58df9067e485&ei=QwBLUsDfAZGo0wXP_4HABQ&usg=AFQj CNEwxMcdxA2p0uYFIwb1byYQolyWrg.

6. See, for example, Supreme Court of the State of New York, *Basis Yield Alpha Fund, Plaintiff, v. Goldman Sachs Group, Inc.,* www.pf2se.com/pdfs/LitigationCases/basis%20v%20goldman.pdf. See also SEC press release, July 15, 2010, "Goldman Sachs to Pay Record $550 Million to Settle SEC Charges Related to Subprime Mortgage CDO," available at http://sec.gove/news/press/2010/2010–123.htm.

7. Goldman Sachs, "Risk Management and the Residential Mortgage Market," http://online.wsj.com/public/resources/documents/goldman0424.pdf.

8. US Senate, Permanent Subcommittee on Investigations, "Wall Street and the Financial Crisis," Hearings, April 27, 2010, testimony of Lloyd C. Blankfein, 134, 135.

9. US Senate, Permanent Subcommittee on Investigations, "Failing to Manage Conflicts of Interest," 398, 404.

10. Ibid., 404, 405–406.

11. Ibid., 408.

12. Ibid., 412.

13. Ibid., 475.

14. Senate Permanent Subcommittee on Investigations, "Wall Street and the Financial Crisis," Hearings, April 27, 2010, 134.

15. Ibid., 137.

16. Ibid., 139.

17. US Senate, Permanent Subcommittee on Investigations, "Failing to Manage Conflicts of Interest," 541–559.

18. Ibid., 549–552.

19. Complaint, *Basis Yield Alpha Fund v. The Goldman Sachs Group, Inc.,* filed October 27, 2001, in the Supreme Court of the State of New York, County of New York.

20. US Senate, Permanent Subcommittee on Investigations, "Failing to Manage Conflicts of Interest," 412.

21. Ibid., 543, 535–536.

22. Ibid., 543.

23. Ibid.

24. Ibid., 544.

25. Ibid., 551.

26. Ibid., 559.

27. Ibid., 545.

28. Ibid., 546.

29. Ibid., 547, 545.

30. Ibid., 552. See also Supreme Court of the State of New York, *Basis Yield Alpha Fund, Plaintiff, v. Goldman Sachs Group, Inc.*, 23–24.

31. Complaint, *Basis Yield Alpha Fund v. The Goldman Sachs Group, Inc.*, filed October 27, 2001, in the Supreme Court of the State of New York, County of New York.

32. Susan Craig and Ben Protess, "Former Trader Is Found Liable in Fraud Case," *New York Times*, August 1, 2013.

Notes to Chapter 6: Emotional Labor

1. Arlie Russell Hochschild, *The Managed Heart: Commercialization of Human Feeling* (Berkeley: University of California Press, 2003), 3.

2. Ibid., 7.

3. Michelle K. Duffy et al., "A Time Based Perspective on Emotional Labor Performance," in vol. 29 of *Research in Personnel and Human Resources Management*, edited by Hui Liao, Joseph J. Martocchio, and Aparna Joshi (Bingley, UK: Emerald, 2010), 96.

4. For the original spoof, see Alan D. Sokal, "Transgressing the Boundaries: Towards a Transformative Hermeneutics of Quantum Gravity," *Social Text* 46–47, nos. 1–2 (1996), www.physics.nyu.edu/sokal/transgress_v2/transgress_v2_singlefile.html. See also Alan D. Sokal, "Transgressing the Boundaries: An Afterword," www.physics.nyu.edu/sokal/afterword_v1a/afterword_v1a_singlefile.html; and the further discussion of the hoax in Steven Weinberg, "Sokal's Hoax," *New York Review of Books*, August 1996, www.nybooks.com/articles/archives/1996/aug/08/sokals-hoax/?pagination=false.

5. Hochschild, *Managed Heart*, esp. chap. 6.

6. Ibid., 127.

7. Alicia A. Grandey, "Emotion Regulation in the Workplace: A New Way to Conceptualize Emotional Labor," *Journal of Occupational Health Psychology* 5, no. 1 (2000): 98.

8. J. J. Gross, "The Emerging Field of Emotion Regulation: An Integrative Review," *Review of General Psychology* 2, no. 3 (1998); J. J. Gross and

O. P. John, "Mapping the Domain of Expressivity: Multimethod Evidence for a Hierarchical Model," *Journal of Personality and Social Psychology* 74, no. 1 (1998). See also J. J. Gross and R. A. Thompson, "Emotion Regulation: Conceptual Foundations," http://med.stanford.edu/nbc/articles/4%20-%20 Emotion%20Regulation%20-%20Conceptual%20Foundations.pdf, 22.

9. Quoted in Grandey, "Emotion Regulation in the Workplace," 97.

10. See, for example, Duffy et al., "Time Based Perspective," 4.

11. Gross, "Emerging Field of Emotion Regulation," quoted in Grandey, "Emotion Regulation in the Workplace," 98. See also Gross and Thompson, "Emotion Regulation: Conceptual Foundations."

12. Grandey, "Emotion Regulation in the Workplace," 99.

13. Ibid., 103–104.

14. Ibid., 105.

15. For example, see Duffy et al., "Time Based Perspective."

16. A. Parasuraman, Valarie A. Zeithaml, and Leonard L. Berry, "A Conceptual Model of Service Quality and Its Implications for Future Research," *Journal of Marketing* 49 (Fall 1985): 47.

17. Gilbert Ryle, *The Concept of Mind* (New York and London: Routledge, 2009).

18. Thomas A. Kochan, "The Social Legitimacy of the HRM Profession: A US Perspective," in *The Oxford Handbook of Human Resource Management,* edited by Peter Boxall, John Purcell, and Patrick Wright (Oxford: Oxford University Press, 2008), 600–601.

19. See Barack Obama, speech at Osawatomie, Kansas, December 6, 2011, www.whitehouse.gov/the-press-office/2011/12/06/remarks-president-economy-osawatomie-kansas; and Tamar Lewin, "Community Colleges Cutting Back on Open Access," *New York Times,* June 23, 2010, www.ny times.com/2010/06/24/education/24community.html?pagewanted=all.

20. For youth unemployment rates in Germany, Spain, and Greece as of December 2012, see Eurostat, news release, "EuroIndicators," February 1, 2013, http://epp.eurostat.ec.europa.eu/cache/ITY_PUBLIC/3-01022013 -BP/EN/3-01022013-BP-EN.PD. For Italy, see Y Charts, "Italy Youth Unemployment for December 2012," http://ycharts.com/indicators/italy _youth_unemployment_rate_lfs.

Notes to Chapter 7: The Military Half

1. Harold J. Leavitt and Thomas L. Whisler, "Management in the 1980s," *Harvard Business Review,* November–December 1958, 44.

2. Ibid., 42, 43.

3. Ibid., 47, 41.

4. On the teething problems of early civilian CBSs, see Thomas Haigh, "Inventing Information Systems: The Systems Men and the Computer, 1950–1968," *Business History Review 75* (Spring 2001); and Thomas Haigh, "How the Computer Became Information Technology: Constructing Information in Corporate America, 1950–2000," thaigh@acm.org.tomand maria.com/.

5. Haigh, "Inventing Information Systems," 15, 16.

6. The main sources for this account of the role of information systems in the Battle of the Atlantic are Clay Blair, *Hitler's U Boat War: The Hunted, 1942-1945* (New York: Random House, 1998); Richard Overy, *Why the Allies Won* (London: Jonathan Cape, 1995), 46–61; and Winston S. Churchill, *The Second World War,* vol. 2, *Their Finest Hour,* 525–537; vol. 3, *The Grand Alliance,* 98–137; vol. 4, *The Hinge of Fate,* 228–247; vol. 5, *Closing the Ring,* 3–15, 228–246.

7. Churchill, *The Second World War,* vol. 3, *The Grand Alliance,* 135–137.

8. The main sources for this account of SAGE and the development of military-based information systems post–World War II are Agatha Hughes and Thomas P. Hughes, eds., *Systems, Experts, and Computers: The Systems Approach in Management and Engineering, World War II and After* (Cambridge: MIT Press, 2000), 1–26; Eric P. Rau, "The Adoption of Operations Research in the United States During World War II," chap. 2 in ibid.; Thomas P. Hughes, *Rescuing Prometheus* (New York: Vintage Books, 2000), introduction and chap. 2; and Paul N. Edwards, *The Closed World: Computers and the Politics of Discourse in Cold War America* (Cambridge: MIT Press, 1997), chaps. 2–3.

9. T. Hughes, *Rescuing Prometheus,* 52–54.

10. Haigh, "Inventing Information Systems," 40.

Notes to Chapter 8: The Nuclear Half

1. Laurence Freedman, *The Evolution of Nuclear Strategy* (London: Macmillan, 1981), 156.

2. Robert Scheer, *With Enough Shovels: Reagan, Bush, and Nuclear War* (London: Secker and Warburg), 13.

3. Ibid., 217–218.

4. Ibid., 277.

5. Paul N. Edwards, *The Closed World: Computers and the Politics of Discourse in Cold War America* (Cambridge: MIT Press, 1997).

6. Scheer, *With Enough Shovels*, 250.

7. On Mormon opposition to the MX, see "Anti-MX Missile Stand Surprised Some Mormons Too," *Salt Lake City Tribune,* May 2, 2011, www.sltrib.com/sltrib/news/51716448-78/church-mormons-utah-statement.html.csp. For Senator Laxalt's opposition, see William E. Schmidt, "Opposition to MX Missile Gaining in Utah and Nevada," *New York Times,* June 8, 1981, www.nytimes.com/1981/06/08/us/mx-opposition-gaining-in-utah-and-nevada.html.

8. For the full text of Reagan's Star Wars speech, see http://pierretristam.com/Bobst/library/wf-241.htm.

9. Lou Cannon, *President Reagan: The Role of a Lifetime* (New York: Simon and Schuster, 1991), 333.

10. Leaked to the *New York Times* and quoted in Freedman, *Evolution of Nuclear Strategy,* 406.

11. For a discussion of Sakharov's role, see Frances FitzGerald, *Way Out There in the Blue: Reagan, Stars Wars, and the End of the Cold War* (New York: Simon and Schuster, 2000), 410–411.

12. See, for example, Mikhail Gorbachev, "A Farewell to Nuclear Arms," www.project-syndicate.org/commentary/a-farewell-to-nuclear-arms.

13. For a detailed account of these disputes, see FitzGerald, "Hardliners vs. Pragmatists," chap. 7 in *Way Out There in the Blue,* 265–313. See also George Schultz, *Turmoil and Triumph: My Years as Secretary of State* (New York: Scribner's 1993), esp. chap. 17, "The Strategic Defense Initiative."

14. Edmund Morris, *Dutch: A Memoir of Ronald Reagan* (New York: Random House, 1999).

Notes to Chapter 9: The Chinese Model

1. Josephine Moulds, "China's Economy to Overtake US in Next Four Years, Says OECD," *Guardian,* November 9, 2012.

2. Eva Dou, "China Gains Tablet-Computer Foothold," *Wall Street Journal Europe,* March 15, 2013.

3. See, for example, Shigeo Shingo, *A Revolution in Manufacturing: The SMED System* (Portland, OR: Productivity Press, 1990). Shingo was one of the chief architects of the Toyota system of manufacture. In his curriculum vitae at the end of the book, Shingo refers to the moment in 1931 when he read Taylor's *Principles of Scientific Management* "and decided to make the study and practice of scientific management his life's work."

4. Xiangming Chen, "A Tale of Two Regions in China: Rapid Economic Development and Slow Industrial Upgrading in the Pearl River and the Yangtze River Deltas," *International Journal of Comparative Sociology* 48 (2007): 174; Ching Kwan Lee, "The Making of New Labor in the Sunbelt," in *Working in China: Ethnographies of Labor and Workplace Transformation,* edited by Ching Kwan Lee (London and New York: Routledge, 2007), 161.

5. Chun Yang, "Strategic Coupling of Regional Development in Global Production Networks: Redistribution of Taiwanese Personal Computer Investment from the Pearl River Delta to the Yangtze River Delta, China," *Regional Studies* 43, no. 3 (2008): 385–407.

6. Ibid., 391.

7. Ibid., 397–399.

8. Ibid., 401.

9. Students and Scholars Against Corporate Misbehaviour (SACOM), "Workers as Machines: Military Management in Foxconn," October 12, 2010, http://sacom.hk/workers-as-machines-military-management -in-foxconn/, 1–9.

10. Jackie Pullinger, *Crack in the Wall: The Life and Death of Kowloon Walled City* (London: Hodder and Stoughton, 1993).

11. Multiple photographs of Shenzen electronics plants can also be accessed through "Electronics Plants, Foxconn, Shenzen," Google Images.

12. Ching Kwan Lee, "Engendering the Worlds of Labor: Women Workers, Labor Markets, and Production Politics in the South China Economic Miracle," *American Sociological Review* 60, no. 3 (1995): 5–6.

13. Mike Parker and Jane Slaughter, "Management by Stress: The Team Concept in the US Auto Industry," www.tandfonline.com/doi/pdf/10.1080/09505439009526271.

14. SACOM, "Workers as Machines," 23.

15. Ibid., 3, 10, 14.

16. Ching Kwan Lee and Eli Friedman, "Remaking the World of Chinese Labour: A 30-Year Retrospective," *British Journal of Industrial Relations* 48, no. 3 (2010): 510.

17. See Pun Ngai, "Gendering the Dormitory Labor System: Production, Reproduction, and Migrant Labor in South China," *Feminist Economics* 13, nos. 3–4 (2007): 243.

18. Lee and Friedman, "Remaking the World of Chinese Labour," 512.

19. Lee, "Making of New Labor in the Sunbelt," 169.

20. See, for example, Zhe Zhang et al., "A Framework for ERP Systems Implementation Success in China: An Empirical Study," *International Journal of Production Economics* 98 (2005); and Kai Reimers, "Implementing ERP Systems in China," *Communications of the Association for Information Systems* 11 (2003).

21. Kimberly Chong, "The Work of Financialization: An Ethnography of a Global Management Consultancy in Post-Mao China" (D.Phil. thesis, London School of Economics, 2012).

22. Ibid., 160, 162.

23. Ibid., 27.

24. Ibid., 164, 166.

25. David Norton, Balanced Scorecard Collaborative, and SEM Product Management, SAP AG, *SAP Strategic Enterprise Management, Translating Strategy into Action: The Balanced Scorecard,* 1999, www.hochschul-management.de/BSC-SAPSEM.pdf.

26. Ibid., 20.

27. Ron Amann, "A Sovietological View of Modern Britain," *Political Quarterly* 74, no. 4 (2003).

28. Ibid., 470.

29. Chong, "Work of Financialization," 84–85.

Notes to Chapter 10: Any Way Out?

1. Lucy Suchman, *Plans and Situated Actions: The Problem of Human-Machine Communication* (Cambridge: Cambridge University Press, 1987); Lucy Suchman, "Office Procedure as Practical Action: Models of Work and System Design," *ACM Transactions on Office Information Systems* 1, no. 4 (1983): 320–328; Lucy Suchman, "Making Work Visible," *Communications of the ACM* 38, no. 9 (1995): 56–64, esp. 60; Claus Otto Scharmer, "Conversation with Lucy Suchman, Xerox PARC," August 19, 1999, www.iwp.jku.at/born/mpwfst/02/www.dialogonleadership.org/Suchman x1999.html.

2. See especially Jacob Hacker, *The Great Risk Shift and the Decline of the American Dream* (Oxford: Oxford University Press, 2008).

3. For an account of employee-owned businesses worldwide, including the Mondragon Cooperatives and the John Lewis Partnership, see Robert Oakeshott, *Jobs and Fairness: The Logic and Fairness of Employee Ownership* (Norwich: Michael Russell, 2000).

4. See Clement J. McDonald et al., "The Indiana Network for Patient Care: A Working Local Health Information Infrastructure," *Health Affairs* 24, no. 5 (2005), www.regenstrief.org/Members/cmcdonald/cmcdonald_bibliography/McDonald2005a.

5. Jon Bakija and Bradley T. Heim, "Jobs and Income Growth of the Top Earners and the Causes of Changing Income Inequality: Evidence from US Tax Return Data," working paper, Williams College, Office of Tax Analysis, March 17, 2009, table 1, quoted in Jacob S. Hacker and Paul Pierson, *Winner-Take-All Politics: How Washington Made the Rich Richer and Turned Its Back on the Middle Class* (New York: Simon and Schuster, 2010), 46.

6. Thomas Frank, *What's the Matter with Kansas? How Conservatives Won the Heart of America* (New York: Holt Paperbacks, 2005).

ACKNOWLEDGMENTS

In writing and doing research for this book, I've had the privilege and good fortune of being associated with two great universities on each side of the Atlantic: Oxford, where I was a senior and associate fellow at the Rothermere American Institute, 2005–2011, and where I have been a senior member at St. Antony's College since 2011; and the Institute for Public Knowledge (IPK) at New York University, where I have been a fellow since 2010. At the Rothermere Institute I am especially grateful to those who made it a stimulating place to work and do research, especially Andrea Beighton, Jim Bell, Peter Bourne, David Ellwood, Matthew Feldman, Godfrey Hodgson, Cheryl Hudson, Mary King, Laura Lauer, and Jane Paulson, librarian at the Vere Harmsworth Library.

I have also benefited greatly from being associated with St. Antony's College, the beacon of internationalism at Oxford, and I am very grateful to Margaret MacMillan, warden of St. Antony's, for making this possible. Another of the great privileges of Oxford is the use of its magnificent libraries, and I would like to thank Sarah Thomas, Bodley's librarian, and her staff, and also Janet McMullin, senior assistant librarian at Christ Church Library.

I would also like to thank my Oxford colleagues at the Council for the Defence of British Universities—Howard Hotson at St. Anne's College, Tim Horder at Jesus College, and Keith Thomas at All Souls College—for their insights into the ways in which some of the practices described in this book have impacted the British universities. At Oxford I am also grateful for the support of Christopher Lewis, dean of Christ Church, and Rhona Lewis; Avner Offer and Peter Pulzer at All Souls College; Paul Flather at Mansfield College and the Europaeum; Bill Dutton at the Oxford Internet Institute; Paola Mattei at St. Antony's; and Bent Flybjerg at the Said School of Business.

At the Institute for Public Knowledge at New York University, I have benefited greatly from the advice and support of Craig Calhoun, formerly director of the IPK and now director of the London School of Economics. The IPK has also been an ideal place to do research, and I would thank all those who have made this possible, especially Eric Klinenberg, Craig Calhoun's successor as director of the IPK, as well as Sam Carter, Jessica Coffey, Martha Poon, Harel Schapira, Richard Sennett, Kimberly Chong, and Caitlin Zaloom. I would also like to thank Bruce Buchanan, director of the Business and Society Program at the Stern School of Business at New York University, for making possible the public lectures I gave at Stern in April–May 2013 on the theme "Towards a New Industrial State."

As director of programs at the New York Review of Books Foundation, I have been able to help organize conferences on both sides of the Atlantic dealing with some of the themes of this book. At the *Review* I would especially like to thank Robert Silvers, editor of the *New York Review,* and Rea Hederman, publisher, without whose generous and enthusiastic support this could not have happened; and also at the *Review* Hugh Eakin, Angela Hederman, Matthew Howard, Mike King, Nancy Ng, Diane Seltzer, Raymond Shapiro, and Catherine Tice. In the planning of these conferences, I've also worked with the Fritt-Ord Foundation of Oslo, among their most generous benefactors, and I would particularly like to thank Grete Brochmann, vice chair of the Board of Trustees at Fritt-Ord; Erik Rudeng, director of Fritt-Ord; and Bente Roalsvig, director for projects.

I owe a great debt of gratitude to my agent, Zoë Pagnamenta, who has been a wise, reassuring, and extremely effective presence throughout the life of this project. Also to Tim Bartlett, my editor at Basic Books. Tim was my editor at Oxford University Press for my last book, *The New Ruthless*

Economy: Work and Power in the Digital Age, and it has been providential that our paths have come together again at Basic. Tim is a great editor who combines enthusiasm and encouragement with an unerring sense of what works and what doesn't, and of what needs to go and what needs to be added. At Basic I'm also grateful for the support of Kate Bowman, Katy O'Donnell, Collin Tracy, Annette Wenda, and Kaitlyn Zafonte.

I would like also to acknowledge the generosity and encouragement of family and friends on both sides of the Atlantic, who have often had to listen to stories of arcane IT systems, but who have often contributed stories of their own. So my brother and sister-in-law, Richard and Alicia Head, and my sister, Tessa Haddon. Also Dee Aldrich, Nelson Aldrich, Elizabeth Baker, Annabel Bartlett, Bill Bradley, Peter Carey, Hugh Cecil, Katherine Cecil, Mirabel Cecil, Laurence Cockcroft, Caroline Egremont, Max Egremont, Ed Epstein, Frances FitzGerald, Grey Gowrie, Neiti Gowrie, Alexandra Howard, Philip Howard, James Howard-Johnston, Robin Lane-Fox, Anthony McCall, David Merrill, Biddy Passmore, Adam Ridley, Robert Skidelsky, Jim Sterba, Sigrun Svavarsdøttir, and Marc Trachtenberg.

Oxford and New York, October 2013

INDEX